DICK KNOWLES obtained the degree of Master of Arts at the University of Nottingham after gaining his first degree in geography at the University of Cambridge. He worked for some years as head of the Department of Geography at Rickmansworth Grammar School, and has also worked as an extra-mural tutor for the University of London. He is now a senior lecturer in the Department of Geography at the Polytechnic of North London. He is also involved with examining at school level, and is the author of a number of publications, among them the *Geography* and *Regional Geography* cards in the Key Facts series and, as co-author, *Europe in Maps*, Vols. I and II, *North America in Maps* (Longman), *Economic and Social Geography* (W. H. Allen) and *Spotlight Project Cards* in geography (Collins).

key facts

GCE O-Level Passbooks

CHEMISTRY, C. W. Lapham, M.Sc., A.R.I.C.

MODERN MATHEMATICS, A. J. Sly, B.A.

HISTORY (*Social and Economic*, 1815–1939), M. C. James, B.A.

FRENCH, G. Butler, B.A.

ENGLISH LANGUAGE, Robert L. Wilson, M.A.

BIOLOGY, R. Whitaker, B.Sc. and J. M. Kelly, B.Sc., M.I.Biol.

PHYSICS, B. P. Brindle, B.Sc.

GCE O-Level Passbook
Geography

R. Knowles, M.A.

Published by Intercontinental Book Productions
in conjunction with Seymour Press Ltd.
Distributed by Seymour Press Ltd.,
334 Brixton Road, London, SW9 7AG

Published 1976 by Intercontinental Book Productions,
Berkshire House, Queen Street, Maidenhead, Berks., SL6 1NF
in conjunction with Seymour Press Ltd.

1st edition, 1st impression 2.76.0
Copyright © 1976 Intercontinental Book Productions

Made and printed by C. Nicholls & Company Ltd

ISBN 0 85047 900 2

Contents

Introduction

This book has been written with the needs of GCE O-level and CSE examination candidates closely in mind. In planning the contents of the book a careful study was made of the syllabuses and recent examination papers of all O-level examining boards and a wide selection of CSE boards. Obviously, in a book of this size it has not been possible to include every topic set by every examining board, but the book does represent the essential core of material required for both types of geography examination. Examination candidates are advised to check the requirements of their own syllabus. Certain topics included here may not be needed for your particular examination; conversely, a few topics set by one or two boards only have not been covered in this book. In particular, it should be noted that no attempt has been made to include any regional geography, owing to the multiplicity of regions prescribed for study, and the fact that these set regions usually change from year to year. Information on these latter topics can be obtained by reference to the appropriate recommended textbooks listed in the Examination Hints on page 187.

With the exception of Chapter 1, metric measurements have been used throughout the book. Thus, distances are expressed in kilometres, heights in metres, and areas in hectares. Temperature figures are given in degrees centigrade, and rainfall totals in millimetres. Production figures are given in tonnes (short tons). Students are advised to become familiar with these metric units and to use them at all times in their preparation for the examination. Chapter 1 deals with map reading, which for most examining boards still involves the use of OS maps at non-metric scales (1 inch : 1 mile, etc.) with heights in feet. Thus, in Chapter 1 the use of non-metric units was considered more appropriate.

In the various chapters dealing with topics in human geography all production and population figures relate to 1972, unless otherwise stated. These were the latest figures available at the time of writing, but students are advised to keep their statistical information as up-to-date as possible and to amend these figures as new production and population statistics become available.

Chapter 1
Map Reading

All GCE O-level and CSE examinations in geography include questions on map reading, and quite clearly this is an extremely important area of study for any geography student. An ability to present information in cartographic or map form and to interpret maps correctly is one of the basic skills of any geographer. Mapwork forms a compulsory section in most O-level and CSE examinations papers, frequently carrying a high proportion of the total marks for the paper on which it appears. Questions are normally set on the 1 inch : 1 mile (1:63,360) or 2½ inch : 1 mile (1:25,000) series of the Ordnance Survey (OS), although with the recent introduction by the Ordnance Survey of a metric scale at 1:50,000 a few of the examining boards also include the new series on their syllabus. Make sure you know which map series are included on your own syllabus.

It is important to gain practice in map reading, and for this purpose copies of OS map sheets should be obtained and carefully studied. Learn the **conventional signs** used on both the 1 inch : 1 mile and 2½ inch : 1 mile maps. (The symbols employed on the new metric scale maps are essentially the same as those on the 1 inch : 1 mile maps, but drawn at a slightly larger size.) Examinations frequently include simple questions designed to test the student's understanding of the meaning of a selection of conventional signs. The commonest mistakes are concerned with the meanings of dashed and dotted lines (various boundaries, footpaths, roads without hedges or fences, etc.) and also with shadings which indicate railway embankments and cuttings.

Techniques of map reading

Grid references

All OS maps are covered by a series of 1 kilometre grid squares which form part of the national grid system. This grid system enables the location of any point on an OS map to be expressed by means of a grid number. The map reference for any point will be the same irrespective of the scale of the map being used.

Map references are needed both in questions directly requiring them and in descriptive questions in which examples have to be quoted. Quoting them wrongly is a very common lapse and can lose marks. Remember that the easting figure (found along the top or bottom of the map) is given before the northing figure (found along the sides of the map). For larger features such as a wood, a lake or a village, it is usually sufficient to quote a four-figure reference (the two easting figures and the two northing figures). For small features such as a church or a spot height, it is necessary to give a six-figure reference. This involves estimating tenths of a kilometre square from the easting and northing figures to give the third and sixth figures of a full six-figure reference. A diagram explaining the method of using the national grid is given in the margin of all OS maps.

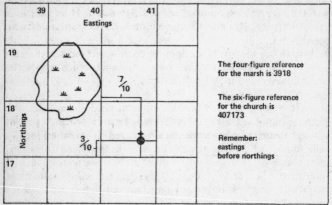

The four-figure reference for the marsh is 3918

The six-figure reference for the church is 407173

Remember: eastings before northings

Figure 1. Grid references

Measurement of distances

The measurement of distances along winding routes such as roads, rivers or railways can be done by various methods: first, by stepping off the distance with a pair of dividers set a known short distance apart; secondly, by marking off the route along the straight edge of a piece of paper, following the bends with the paper but being careful not to move it sideways; and thirdly, by following the route with a piece of cotton or very thin string. Whichever method is employed, the map distance must then be converted to a ground distance by means of the scale-line on the map. The result may be expressed in either miles or kilometres or fractions of these units. If a distance is required in kilometres it is sometimes useful to remember that the grid lines provide a scale since they are placed at 1 kilometre intervals.

Examination candidates frequently lose marks on this type of question owing to inaccurate measurement. Whichever method is used, it is most important to check the result by repeating the measurement in the opposite direction along the route.

Calculation of average gradient

For the calculation of the average gradient between two points it is necessary to establish from the map the horizontal distance between the two points and also the difference in height between them. Both measurements must be reduced to the same units — say feet or metres. The average gradient is then calculated by dividing the horizontal distance by the vertical difference, and expressing the result as 1 in x. If we assume that the horizontal distance between two points is $2 \cdot 3$ miles and that one point is 171 feet higher than the other, then the calculation would be as follows:

$$\text{horizontal distance} = 2 \cdot 3 \times 5280 \text{ feet}$$
$$\text{difference in height} = 171 \text{ feet.}$$
$$\text{Average gradient} = \frac{2 \cdot 3 \times 5280}{171}$$
$$= 1 \text{ in } 71$$

Notice that the gradient is expressed as 1 in 71, **not** as 1 foot in 71 feet. Remember too that the full working must be clearly shown.

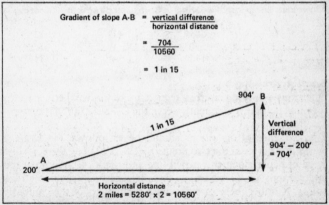

Figure 2. Calculation of gradient

Occasionally a question is set which involves calculating the steepest slope in a given area. This is obtained by finding the point at which any two successive contours are at their closest distance apart. Assuming the map to be on a scale of $2\frac{1}{2}$ inch : 1 mile, then the difference in height between contours will be 25 feet. Thus, if the closest pair of successive contours are 1/10 inch apart, the calculation will be:

land falls 25 feet in 1/10 inch on map;
land falls 25 feet in $1/10 \times 2/5$ miles on ground;
land falls 25 feet in $1/10 \times 2/5 \times 5280$ feet on ground;
land falls 1 foot in $1/10 \times 2/5 \times 5280 \times 1/25$ feet on ground;
land falls 1 foot in 8·4 feet on ground.
$\therefore$ steepest gradient is 1 in 8.

Measurement of bearing

The most frequent mistake in this type of question is to confuse bearing with compass direction. A bearing is an angular measurement taken in a clockwise direction from north; a compass direction is described according to the thirty-two cardinal compass directions – NE, SSE, SW, etc. For the purposes of O-level work, 'north' is taken to be 'grid north'. Thus, to find a bearing, draw a line between the two points and a north-south line through the point from which the bearing is to be measured. The north-south line should be drawn parallel to the north-south grid lines marked on the map. Place a protractor on the point from which the bearing is to be measured and read off the angle from north. Full circle angular bearings for the main compass directions are listed below.

Compass direction	Bearing	Compass direction	Bearing
N	000° or 360°	S	180°
NE	045°	SW	225°
E	090°	W	270°
SE	135°	NW	315°

Section drawing

Questions requiring a cross-section to be drawn between two specified points are frequently set in the O-level examination.

The sequence of operations should be as follows. (1) Draw a pencil line between the two points specified. (2) Lay the edge of a piece of paper along this pencil line and mark off accurately the intersections of all contours, spot heights and trigonometrical points. Number these carefully with the heights of the contours, etc. (Students occasionally find this difficult because of the lack of height indication on the contour lines. If the map is studied carefully, however, it is always possible to fix some of the heights, and judgement of slopes will enable the others to be fitted in.) (3) Take a piece of graph paper and draw a base line for your section equal in length to the distance between the points specified. Draw perpendicular lines from either end of this line and mark off the vertical heights. Poor choice of vertical scale is frequently a reason for marks being lost. If the map is on a scale of $2\frac{1}{2}$ inch : 1 mile, contours are at 25 foot intervals and the correct vertical scale is 25 feet to 1/10 inch. If the map's scale is 1 inch : 1 mile, then 50 feet to 1/10 inch should be used. (4) Lay the piece of paper showing heights along the base line (0 feet) on the graph paper and transfer these heights to the appropriate level on the graph paper. (5) Draw in the section carefully with curved lines; the landscape should not appear angular.

In some questions the problem of **intervisibility** is involved. In this case a straight line of sight should be drawn on the section between the two points specified. Whether or not this line intersects any of the land surface will prove the intervisibility of the two points. Some questions will involve annotating the section with names of the physical features or settlements. These should be printed neatly above the section and vertical arrows drawn to indicate the appropriate parts of the land surface.

Occasionally a calculation of the **vertical exaggeration of scale** is required for your section. This is simply a comparison of the horizontal and vertical scales employed. An example might be worked as follows:

vertical scale (VS) = 50 feet on map: 1/10 inch on section;
horizontal scale (HS) = 1 mile on map: 1 inch on section.
Thus, the VS = 1 inch: 500 feet: HS = 1 inch: 5280 feet.
Thus, the vertical exaggeration is $\frac{5280}{500} = 10 \cdot 5$ times.

Measurement of areas

Questions involving the measurement of areas are not commonly set since they take rather a long time. The method employed on a 1 inch : 1 mile map with an irregular area such as a lake or wood is to trace off the outline of the feature and transfer it to 1/10 inch square graph paper. The number of 1/10 inch squares completely within the outline, together with the number which lies half or more inside the outline, is then totalled. Since each 1/10 inch square represents 1/100 square mile on the ground, the area can be found. Thus, if there are 79 squares half or more within the outline and 56 lying wholly within it, the area will be $79 + 56 = 135 \div 100 = 1 \cdot 35$ square miles.

Map reading: description

As well as questions which test the techniques of measurement from maps, most O-level and CSE examinations include questions which require certain features or sections of a map sheet to be described. Such descriptions must be written in an orderly manner. The chief mistake candidates make is to present a haphazard and disconnected series of observations. The following series of points should serve as a guide to the organisation of some of the more commonly required types of description.

Description of upland areas

Comment on the average height of the upland and mention the maximum height attained. Mention the steepness of slopes, the existence of any passes or cols through the upland. Comment on the degree to which the upland is dissected by river valleys and the pattern which they follow (radial, dendritic or trellis, etc.). Conclude your description by attempting a classification of the upland (plateau, dissected plateau, glaciated highland, scarpland, etc.), and mention any peculiarity of rock type which may be apparent (limestone or chalk country, resistant and impermeable rock, etc.).

Description of a valley

Examine the course of the valley. Is it winding or straight? Are there many interlocking spurs? Does the river flow in a wide, straight and open valley, or a narrow valley with high land on either side? Examine the cross-section of the valley at various points along the river's course. Comment on the steepness of the sides, the width of the flat valley floor, if any. Examine the downstream gradient of the river. Are there few or many contours crossing the river? Estimate the approximate gradient. Try to conclude with some generalisation as to the type of river valley. Is it youthful, mature or old age? Does it show the characteristics of a glaciated valley? Has it the abnormal characteristics of valleys in limestones areas?

Description of a section of coast

Mention the direction of the coast. Is the coast straight, a series of open bays and headlands, or is it cut into by a series of narrow inlets? If the coast is of more than one type, subdivide and describe each in turn. Look for depositional features (wide sandy beaches, lines of dunes, spits, bay bars, etc.), or erosional features (cliffs, wavecut platform, caves, arches, stacks, etc.). Comment on the size of features, e.g. height of cliffs, width of beach between high and low water mark.

View from a point

Occasionally a question is set which involves the description of the view from a hilltop or other specific point. This should also be tackled in a logical manner. If the question mentions the direction of view then one can assume that the countryside for 15° on either side of the line can be seen. Descriptions should follow one of two patterns: either describe features in the foreground, then the middle distance and finally in the distance; or describe what can be seen on the left of the view then the centre and lastly on the right-hand side. Comments on the type and size of features should be made as outlined above.

A complete physical analysis

Another type of question is that which involves the subdivision and description of either part or the whole of a map sheet or map extract. Subdivision must be made before description. This will usually require a simple sketch map. Before attempting to draw the sketch map, spend some time in examining the map and working out a simple subdivision. It is usually helpful to select one or two contours and draw over them with a thick pencil. Contours which approximately divide upland from lowland or a flat flood plain from undulating country, etc., should be treated in this way. Once the regions have been selected the sketch map can be drawn. This should be kept as simple as possible. Draw the appropriate sized rectangle and insert the courses of the main streams. Draw in the approximate boundaries (usually contours) between the regions and annotate with a brief description of each, e.g. flood plain 150–200 feet above sea level. If a description is required in addition to the sketch map, then write in accordance with the hints outlined above.

Map reading: interpretation

Map interpretation is concerned with the explanation, analysis and relationships between patterns of relief, vegetation, settlement, communications, etc. shown on a map sheet. It is an extension of the simple descriptive type of exercise mentioned previously, and may also involve certain of the skills and techniques of map measurement described earlier. In a sense, therefore, it represents the ultimate test of skill and proficiency in map reading. In analysing patterns of vegetation, land use, settlement, etc., reference should only be made to information and evidence available on the map sheet itself. Additional information or knowledge about the map area should not be introduced. In many cases the map evidence will be suggestive rather than conclusive. For example, it is rarely possible to determine the specific rock type of an area simply on the basis of map evidence, but it should be possible to state that the rock is resistant or impermeable, etc. Similarly, a map can do no more than hint at the reasons for the growth of settlements or the siting of industry. However, it will offer certain clues, and in exercises on map interpretation the aim is to assemble the relevant map information (although it may not provide conclusive evidence) in a clear, logical manner.

Analysis of vegetation patterns

An analysis of the distribution pattern of a particular form of vegetation should focus attention on both the requirements of the vegetation concerned and also the reasons why the land is not used for other more profitable purposes. Thus, when analysing the distribution of woodland or rough pasture, the maximum and minimum heights should be considered along with the relationship to slopes, rock type and water supply. In addition, evidence concerning the quality of soil and the difficulty of access may also help to explain why the land has not been used for crop farming or another more profitable enterprise. The distribution of orchards is often related to minor features of slope and drainage, as well as height above sea level. These may be observable on the map. Other factors, however, such as labour supply and ease of access to a market or a processing plant, may not be so obvious.

Analysis of settlement patterns

Questions relating to the presence or lack of settlement in an area are concerned fundamentally with the possibilities of making a living in that area. Attention should be paid to the following points.

Relief How much land is available for settlement purposes? Are slopes too steep? Is the altitude too great for building and for agriculture?

Drainage Is the land too wet for building or farming activity? (Look for evidence in the form of artificial drainage channels, senile meanders, marshes, etc.) Is there a lack of water? (Settlement in chalk and limestone areas is frequently determined by the presence of springs and shallow wells.)

Soil Does map evidence such as the existence of rock outcrops, rough pasture, marsh, place names, etc., suggest that the soil is of a poor quality and therefore unsuitable for farming?

Communications Is the area isolated or inaccessible by reason of relief, and therefore unsuitable for farming or industries in which goods need to be sent to a market?

Questions on site and position of settlements are frequently answered badly because candidates are not clear about the meanings of the terms. **Site** is the land upon which the settlement is placed and each site has advantages and disadvantages which may be quoted. Thus the answer to a hypothetical question on a village site might be as follows:

'Little Boddington is a long "street" village built on a belt of gently-sloping land between 250 feet and 300 feet above sea level. To the north lies the low flat flood-plain of the River Derwent which from the number of artificial water-courses appears liable to flooding and therefore unsuitable for building. To the south the steeply rising slopes of Boddington Down make building difficult. One of the site advantages appears to be the existence of spring water; the Boddington Beck rises at 101323 just 50 yards south of the centre of the village. The main disadvantage of the site is a lack of space for building. The village has therefore been forced to spread east and west along the road for a distance of over one mile.'

The **position** of a settlement is bound up with its relationship to the surrounding area, a relationship which may have stimulated its growth. Many towns have grown large because they are at a meeting place of routes. It is true to say that the vast majority of towns are route centres. However an analysis of the position of a town must explain why the town is a route centre or nodal point. This will usually involve the analysis of routes and will take into account such features as valleys, gaps or passes, the existence of a narrow part of a valley or river where a bridging point was possible, etc. In the case of a village, analysis of position should include reference to its nearness to land of various types; its activity will be concerned usually with farming rather than trading or manufacturing as in a town. An analysis of the position of a 'village' might proceed as follows:

'Little Boddington is in a suitable position for its inhabitants to engage in a variety of farming activities. To the north the flat plain of the River Derwent provides abundant grass for the pasturing of cattle; to the south the hills of Boddington Down are

17

useful for sheep pasturage, while cereals such as wheat and barley may be cultivated on the better soils. The belt of gently sloping land by the village is an area of intensive farming, both for orchard fruit and other crops. The position of the village on the A402, which runs from east to west between the hills and the plain, enables the village to send its produce to factories and market in the town of Derwent, seven miles to the west.'

Detailed reference to the map should of course be made throughout such analyses.

Analysis of communication patterns

Questions on communications are almost always concerned with the relationship between relief and roads and/or railways. The most effective answer is to draw a simple sketch map and comment briefly on the points brought out on the sketch map. The sketch map should be clearly drawn, but not over-elaborate. Concentrate attention on the main lines of communication ('A'- and 'B'-class roads and railways if required) and ignore minor roads, trackways and footpaths, unless these are unique in that they form the only link across, for example, an escarpment. It may be helpful to outline the areas of highland on the OS map by drawing over a contour with a thick pencil and marking important gaps. Draw a rectangle of suitable size (half scale is usually sufficient). Draw in the main rivers and highland using the contour you have pencilled over as a guide. Shade in areas of flat land which are liable to flood. Mark the chief gaps and bridging points. Insert the main roads and railways with embankments, tunnels and cuttings, together with the chief settlements. It must be emphasised that this need not be a detailed map – speed is essential. Then write a paragraph emphasising the main relationships brought out by your sketch map.

Map and photograph

In recent years a number of the examining boards have made increasing use of aerial photographs in connection with the mapwork questions. These photographs are of two types, vertical' and oblique. In most cases an oblique aerial photograph is presented, taken by a camera pointing obliquely towards the ground. Such a photograph does *not* provide a scaled representation of the terrain, and distance or area measurements cannot be made from the photograph, which usually covers only part of the area shown on the accompanying map sheet. Questions are usually set on one or more of the following problems: measurement of the direction and angle of view of the photograph, orientation of the mapsheet and identification of features on the photograph, and, occasionally, determination of the time of day when the photograph was taken.

The first task is to examine both photograph and map, picking out readily identifiable features such as a village, a hill, a church tower, a wood, etc. Orientate the map so that the various features fall into the same directional relationship as they are upon the photograph. Find two features near the right-hand edge of the photograph and identify them upon the map. Draw a straight line on the map to represent the limit of the photograph on the right side. Repeat for the left side of the photograph. The point at which these two lines intersect is the point from which the photograph was taken. The angle of view is the angle between the two lines upon the map. The direction in which the camera was pointing is given by the line which bisects this angle. It is usual to refer to the compass direction when expressing this. The description of physical features on the photograph should be in an orderly fashion and recourse may be had to the map for additional information on slopes, etc.

To determine the approximate time of day when the photograph was taken, a close examination must be made of the length and orientation of shadows from tall buildings, trees, etc. For example, in the early morning with the sun low in the east, buildings will cast long shadows in a westerly direction.

Key terms

Bearing An angular measurement of one point from another taken in a clockwise direction from north.

Grid system A system of numbered intersecting lines drawn on a map which enable the location of any point to be given by means of a grid reference.

Map scale The ratio between distances on a map and the corresponding distances on the ground. Map scale may be expressed by means of a statement, representative fraction, or scale-line.

Position The location of a village or town in relation to its surrounding area. The situation of a place. Care must be taken to distinguish between the terms 'site' and 'position' when describing settlements.

Site The ground or area upon which a village or town has been built.

Chapter 2
Field Studies

A number of the GCE and CSE examining boards make provision in their syllabus for students to be tested on detailed work carried out in a small field-study area. In some instances the results of such fieldwork must be presented in the form of a short project or field-book (usually of a recommended length of about 2000 words), while in other instances a group of questions are included on the examination paper, designed to test the student's understanding of the geography of his selected area of study.

The main purposes of geographical field study are described by one examining board as follows: firstly, to acquire the habit of accurate observation and, where necessary, to become trained in the use of simple instruments needed for such observations; secondly, to recognise those facts which are of geographical significance, and to learn methods of recording them by means of pictures, maps, graphs and tables in addition to normal description; thirdly, to use such observations and records to build up a knowledge of the associations of facts which form the geography of the selected region; finally, to develop the ability to use this geography for the purpose of envisaging, at least in part, the geographical facts of other regions.

The area selected for study might be the student's school or home district or an area visited during a school journey or vacation. Most examining boards simply state that the field-study area should be 'small', and do not make any specific regulations about a suitable size, although some boards recommend that it should not exceed 100 square kilometres (10 km × 10 km). Fieldwork may be undertaken in either rural or urban areas.

The main topics for investigation might be as follows.
Location and delimitation of the study area Map and comment on the extent and boundaries of the district studied, and note its position in relation to the country as a whole.
Geology, relief and drainage Consult the geological map, and where possible make a first-hand study of rock outcrops and collect rock specimens. Observe and describe the relief and drainage features in the study area. Map and sketch river features, and illustrate the relief with maps and cross-sections.

Weather records Collect data on temperature, rainfall, pressure and winds (see section on weather-recording, page 95).

Soils and vegetation Examine the soils in the study area. Draw maps to show woodland, heathland, commonland, etc. Note the main species of plants and trees, and attempt to relate their distribution to factors such as relief, soil, drainage, etc.

Land use Map the distribution of crops on $2\frac{1}{2}$ inch : 1 mile or 6 inch : 1 mile base maps. Note the types of livestock raised in the area. Try to relate the pattern of land use to climate, relief, drainage, soils, markets and other relevant factors. Detailed sample studies of individual farms (see page 22) can provide valuable information for illustrating crop rotations, the annual work pattern, methods of production and marketing, etc.

Industry Map the distribution of industrial sites in the study area. Attempt a classification of the industrial premises revealed by the survey. Try to relate the different types of industry to raw materials, power sources, labour-supply, communications, markets, etc. Detailed studies of individual factories can provide useful insights into the organisation of industry in the area concerned (see page 23).

Settlement If work is being carried out in a rural area the distribution of different types of settlement such as isolated farms, hamlets and villages should be mapped and analysed. Try to assess the relative importance of the various villages in terms of the services they provide and the size of the surrounding areas which they serve. A study of the building materials used can often be very rewarding. A study of urban settlement is more complicated than one of rural areas. Nevertheless, a carefully organised urban study can yield much useful information. Attention should be directed to the distribution of different urban functions such as housing, retailing, industry, etc. Analysis of street patterns, building types, building materials, etc. can add depth and interest to the study. Historical information should be introduced only if it helps to explain the growth of the settlement or elements of the present settlement pattern.

Communications Produce a series of maps to show the local patterns of roads, railways and canals. Relate these patterns to relief and settlement. Traffic surveys can be conducted to give some indication of the volume, direction and type of traffic utilising the various routes.

Obviously not all of these themes will apply equally to your study area. Emphasise those features which are 'special' to your selected area and which give it its own particular geographical character.

With the exception of the weather observations, maps should be produced for each of the themes relevant to the study area. These should normally be based on 1 inch : 1 mile or $2\frac{1}{2}$ inch : 1 mile scales, depending on the size of the study area, or a 6 inch : 1 mile scale if urban studies are being carried out. Features should be observed and plotted and investigations made into the reasons for the patterns of distribution. A field-notebook should be used, and observations and ideas entered as the work proceeds. Each unit of study should therefore consist of a map, suitably annotated, together with written work analysing the features shown on the map, the latter including diagrams, sketches, photographs, etc. where appropriate.

Many aspects of study in human geography cannot be adequately carried out simply by means of field observations. In the case of investigations into local agriculture or industry, for example, supplementary data should be sought from farmers and factory managers. This will involve preparing a questionnaire which can be completed with the aid of the person concerned. Decide in advance what information is of geographical relevance, prepare your questions carefully, arrange a meeting by telephone or letter, and explain the purpose of your study. Do not expect to get the information you require if you arrive unannounced at a farm or factory with only a vaguely formulated set of questions.

Sample farm study

Much of the data relevant to an individual farm study can be collected by careful observation and recording, but this will generally need to be supplemented by an interview using a prepared questionnaire. The following topics can be incorporated into most farm studies:

Site	— Altitude, aspect, gradient, drainage, etc.
Access	— Distance from main roads and railways.
Soils	— Type, depth, fertilisers used, improvements.
Fields	— Field sizes, boundaries (hedges, fences, walls).
Grazing	— Grass type, rough grazing, permanent pasture.
Crops	— Crop types, yield, rotation, seed sources.
Animals	— Type of livestock, breeding and stocking policy.
Tenure	— Owner occupier or tenant farmer.
Labour	— Size of labour force, casual labour, accommodation.
Machinery	— Implements used.

Organisation	– Daily and seasonal work pattern.
Buildings	– Function, age, materials, improvements.
Water supply	– Piped supplies, ponds, wells, irrigation.
Marketing	– Local and distant markets, government Marketing Boards.
Policy	– Integration, co-operation, subsidies.

Sample factory study

Studies of individual industrial units must be based very largely upon material obtained from interviews and questionnaire surveys. The following themes are among the main items that should be included in such a study.

Site	– Site features, special advantages, if any. Pollution problems.
Position	– Communication links. Accessibility.
Plant layout	– Types of buildings and their arrangement.
Growth	– When was the firm founded? Phases of expansion.
Raw materials	– Raw materials used. Amounts. Methods of transport to factory. Chief sources.
Power	– Chief power sources.
Water supply	– A neglected factor in many industrial studies. Amounts used. Sources.
Production	– Some reference should be made to production processes, but detailed technological accounts alone do not constitute a geographical study.
Labour	– Numbers employed. Nature of the work force (sex, age, special skills, etc.). Place of residence. Journey to work.
Marketing	– Types of products. Volume of production. Methods of transport used for distribution. Chief markets.

Do *not* attempt to answer any questions on field studies or local studies unless you have been involved in organised fieldwork either at school or at a field-study centre. This type of question cannot be adequately answered purely by 'book' information. There is no substitute for collecting information 'in the field' and organising your own material. That is the aim of the field-studies option. Conversely, if you have been able to carry out a field study, make full use of your knowledge. Use your information to provide examples of themes questioned in physical, human and regional geography. Examiners give credit for sample studies and examples based upon personal observations.

Chapter 3
The Planet Earth

The earth has the form of an **oblate spheroid**. That is to say, it is almost perfectly spherical, except for a slight flattening at the poles. This flattening, and an accompanying bulge at the equator, are produced by the centrifugal force of the earth's rotation. The earth has a polar circumference of 40,009 km and an equational circumference of 40,077 km. Its total surface area is approximately 510 million square kilometres.

Latitude and longitude
In order to refer to positions on the earth's surface, a series of imaginary lines are superimposed on the globe. These are lines of latitude and longitude.

The **latitude** of a place is a measure of its angular distance north or south of the equator measured from the centre of the earth. This angle may be determined for a particular place by subtracting the angle of the noon sun above the horizon on the equinox (see page 30) from 90°. Answers to questions requiring an explanation or definition of the term 'latitude' should include a simple diagram showing the earth, and incorporating the plane of the equator, the earth's axis of rotation, and a line joining the centre of the earth and the particular place in question, e.g. London, latitude 51°N.

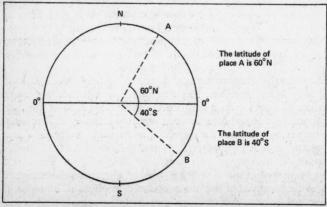

Figure 3. Determination of latitude

Lines of latitude, sometimes referred to as parallels, run east to west, and are drawn parallel to the equator, which is located at an equal distance between the North and South Poles at all points along its length. Lines of latitude (90° in each direction north and south of the equator) are a constant distance apart, approximately 110 km.

The **longitude** of a place is a measure of its angular distance east or west of the **Greenwich meridian** (0°) which runs through Greenwich in London, and has been internationally adopted as the line from which longitude measurements are made. The angle is measured from the axis of the earth. Lines of longitude, sometimes called meridians, run from the North to the South Pole. There are 180 degrees of longitude in each direction east and west of the Greenwich meridian.

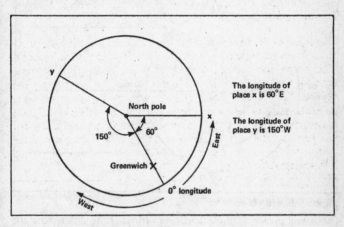

Figure 4. Determination of longitude

These lines of latitude and longitude, as defined above, can be used to give a precise reference for any point in the globe. Each degree of both latitude and longitude is divided into sixty minutes. The form in which these references are given is shown by the three examples below. You will find that the place-name index in your atlas uses the same system.

New York (USA)	40°–45′N,	74°–0′W
Melbourne (Australia)	37°–41′S,	145°–0′E
Tokyo (Japan)	35°–45′N,	139°–45′E

Longitude and time

The **earth's axis** is defined as the shortest diameter joining the North and South Poles (latitudes 90°N and 90°S). The earth rotates from west to east round this axis, completing one rotation or 360° every twenty-four hours. Calculation of time differences between any two places are based on their differences in longitude. Since the earth rotates through 360° in twenty-four hours, it can be calculated that a difference of 15° in longitude represents a time difference of one hour. Since the earth rotates from west to east, it also follows that time is **later** the further east one goes and **earlier** towards the west. An example is given below:

	New Orleans	London	Calcutta
Longitude	90°W	0°	90°E
Local time	6 a.m. (0600 hr)	Noon	6 p.m. (1800 hr)

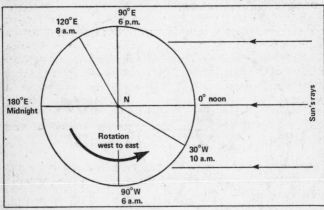

Figure 5. Longitude and time

Questions involving the calculation of local time differences are frequently set in the O-level examination, and students should be familiar with the method of calculation and presentation of the answer. Two examples will serve to illustrate the approach.

Question 1   Calculate the local time at the following places when it is 3 p.m. at Greenwich: (a) Leningrad (30°E), (b) Manaus (60°W), and (c) Tokyo (139°E).

Answer

(a) Difference in longitude between Greenwich (0°) and Leningrad (30°E) = 30°.

Therefore difference in time = 30 ÷ 15 = 2 hours.

Leningrad is **east** of London.

Therefore time in Leningrad is 3 p.m. + 2 hours = 5 p.m.

(b) Difference in longitude is 60° (Manaus) − 0° (Greenwich) = 60°.

Therefore difference in time = 60 ÷ 15 = 4 hours.

Manaus is **west** of London.

Therefore time in Manaus is 3 p.m. − 4 hours = 11 a.m.

(c) Difference in longitude is 139° (Tokyo) − 0° (Greenwich) = 139°.

Therefore difference in time = 139 ÷ 15 = $9\frac{4}{15}$ hours = 9 hours 16 minutes.

Tokyo is **east** of London.

Therefore time in Tokyo is 3 p.m. + 9 hours 16 minutes = 12.16 a.m.

Question 2 What is the local time at San Francisco (122°W) when it is 7 p.m. at Jerusalem (35°E)?

Answer

(This type of question is slightly more difficult since it involves the idea of longitude being calculated both east and west of Greenwich.) San Francisco is 122° west of Greenwich; Jerusalem is 35° east of Greenwich. The difference in longitude is obtained by **adding** the two values. Thus the difference in longitude is 157°. Time difference = 157 ÷ 15 = $10\frac{7}{15}$ hrs. = 10 hours 28 minutes.

San Francisco is **west** of Jerusalem.

Therefore time at San Francisco is 7 p.m. − 10 hours 28 minutes = 8.32 a.m.

Calculations of this type may be facilitated by changing the p.m. time into its 24-hour-clock equivalent. Thus 7 p.m. becomes 1900 hours. 1900 − 1028 = 0832.

Time zones

From the preceding remarks it will be obvious that significant local time differences exist over relatively short distances. For example, two places separated by one degree of longitude will have a difference in local sun time of four minutes. In order to eliminate the confusion that would be created by small time differences over short distances, the world is divided into a number of time zones. These are usually 15° of longitude in extent, and watches must be adjusted by one hour (backwards or forwards, depending on the direction of travel) when passing from one time zone to another. Countries with a large east–west extent may be crossed by a number of time zones. Canada, for example, which stretches through 88° of longitude, from Newfoundland in the east to the Pacific coast in the west, has six time zones.

The International Date Line

Apart from slight deviations around island groups, the International Date Line follows the 180° meridian. When crossing this line adjustments must be made to the date. The explanation for this unusual situation can be best appreciated by drawing a simple diagram. Draw a circle to represent the earth. Mark the centre N to represent the North Pole. Draw a radius and label it 0° (the Greenwich meridian). Draw a further radius at 180° to the 0° meridian and label it 180°. Calculations of time reveal that if worked **east** of Greenwich the local time at 180° becomes twelve hours **later** than Greenwich; if worked **west** of Greenwich, the time at 180° is twelve hours **earlier** than Greenwich. In other words, there is a twenty-four hours difference either side of the 180° meridian. Thus, in crossing the International Date Line towards the west a ship or aircraft loses one whole day; in travelling towards the east across the line a day is repeated.

Great circles

The shortest distance between any two points on the globe lies on a line, the plane of which cuts through the centre of the earth. Such lines are sections of what are termed great circles. It is clear that all lines of longitude are great circles, but that of all the lines of latitude only the equator is a great circle. For example, although Japan and Denmark lie in similar latitudes, the shortest route between them does not lie along a line of latitude but over the North Pole. The simplest, but not completely accurate, diagram to represent these facts is one of the northern hemisphere with the North Pole at its centre. In the example mentioned of Japan and Denmark, the appropriate line of latitude can be shown as a circle, and a dashed line marked to indicate the shortest route between them; namely, the Great Circle route.

For east–west routes in the northern hemisphere a great circle forms an arc north of the appropriate line of latitude; in the southern hemisphere it forms an arc to the south of the line of latitude connecting two points.

The seasons

As well as **rotating** on its axis, the earth is also **revolving** around the sun in a slightly eliptical orbit. The earth, travelling at a speed of more than 96,000 km per hour, takes 365¼ days (a solar year) to complete one revolution, and on average remains about 149 million kilometres from the sun. For convenience a year is taken as 365 days with a leap year of 366 days every fourth year.

The earth's axis of rotation is not perpendicular to its plane of orbit round the sun. If it were, we would have no changing seasons of the year. The axis is, in fact, inclined at an angle $23\frac{1}{2}°$ to the perpendicular or $66\frac{1}{2}°$ to the plane of orbit. The overhead sun therefore appears to move from the Tropic of Cancer ($23\frac{1}{2}°N$) on 21 June to the Tropic of Capricorn ($23\frac{1}{2}°S$) on 22 December. In other words, the two tropics mark the northern and southern limits of the overhead sun.

The higher the sun is in the sky the greater the amount of heat-energy received. Midsummer in the northern hemisphere occurs in June, while in the southern hemisphere this is the period of midwinter. Spring and autumn seasons relate to the intermediate periods between these two extreme positions of the overhead sun.

Length of daylight
The length of daylight experienced at different places each day also varies according to the position of the overhead sun. The northern hemisphere has long nights and short days in December, while the reverse is true in the southern hemisphere at the same time of the year. A comparison of sunrise and sunset times

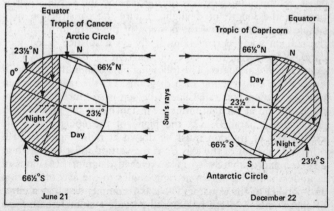

Figure 6. Seasonal variations in length of daylight

(published in many newspapers) for places in northern Scotland and southern England will show how the length of daylight varies according to latitude and season.

Areas north of the Arctic Circle ($66\frac{1}{2}°$N) experience twenty-four hours of daylight on 21 June. This phenomenon is referred to as the **midnight sun**. On the same date areas south of the Antarctic Circle ($66\frac{1}{2}°$S) experience continuous darkness. On 22 December these conditions are reversed (see figure 6).

Solstices and equinoxes

The term 'solstice' is used to refer to the two dates when the sun is overhead at the tropics. 21 June is the **summer solstice** in the northern hemisphere and the date when the sun is overhead at the Tropic of Capricorn. On this date places in the northern hemisphere experience their longest days and shortest nights. 22 December is the **winter solstice** when the sun is overhead at the Tropic of Cancer and places in the northern hemisphere have their longest nights and shortest days.

Similarly, there are two **equinoxes**, 21 March and 23 September. On these dates the sun is overhead at the equator, and all places, irrespective of their latitude, experience twelve hours' daylight and twelve hours' night, with sunrise and sunset at 6.00 a.m. and 6.00 p.m. local sun time respectively.

Most examination questions on the seasons and seasonal variations in the length of daylight require the construction of a simple diagram to show the position of the earth in relation to the sun on a given date (almost always a solstice or equinox date). Very often this type of diagram is badly drawn by examination candidates. The following points should be borne in mind when constructing such diagrams: (1) the rays of the sun must be shown as parallel lines (the sun is so much larger than the earth and at such an immense distance that the rays of light are virtually parallel); (2) the plane of the equator and the earth's axis of rotation must be drawn at right-angles to each other; (3) the tilt of the axis from the vertical must be drawn at $23\frac{1}{2}°$ (or approximately so in a sketched diagram); (4) lines of latitude should be drawn at right-angles to the axis of rotation and parallel to the equator; (5) it is recommended that a circle of approximately 4-cm radius will provide a suitable base for a good, clear diagram.

Determination of position

Occasionally the O-level geography examination includes questions involving simple calculations based on the ideas of longitude and time, and upon the position of the earth in relation to the sun at various seasons. An example is given below:

Question On 23 September a ship's sextant is found to show that the sun is overhead at 12 noon, while its chronometer indicates that the time at Greenwich is 2 p.m. What is the position of the ship? Give your reasons.

Answer 23 September is the autumn equinox. Therefore the sun is overhead at the equator. The longitude of the ship is found by comparing local time on board ship with Greenwich Mean Time. In this case the time difference is two hours, which represents 30° of longitude. The time is earlier on the ship than at Greenwich, therefore the longitude is 30°W. The position of the ship is therefore 0° latitude. 30°W longitude.

Key terms

Great circle An imaginary circle on the earth's surface whose plane passes through the earth's centre. The shortest distance between any two points is along the arc of the Great Circle on which they lie.

Greenwich meridian The line of longitude (0°) which passes through Greenwich in London. This is almost universally used as the standard meridian for measurements of longitude.

International Date Line An imaginary line following the 180° meridian for much of its length. Adjustments are made to the date when this line is crossed.

Latitude The distance of a place north or south of the equator, measured as an angle with the centre of the earth. The equator is latitude 0°, the North Pole is latitude 90°N, and the South Pole is 90°S.

Longitude The distance of a place east or west on the earth's surface, measured by the angle which the meridian of that place makes with some standard meridian. The standard meridian which is almost universally used is that of Greenwich, London, which is regarded as 0°.

Time zone An area within which a standard time is adopted, usually the local time of the meridian running through the centre of the time zone. In large countries, such as the USA, falling into a number of time zones, each is given a name. Thus, Canada has Atlantic, Eastern Standard, Central, Mountain and Pacific Times, respectively 4, 5, 6, 7 and 8 hours behind Greenwich Mean Time.

Chapter 4
Rock Types

The rocks of the earth's crust are extremely varied and may be classified in a number of different ways: for example, according to their mineral composition, age, mode of formation, etc. The most commonly employed system of classification is based on mode of formation, and on this basis three main classes of rocks may be identified: namely, igneous, sedimentary and metamorphic rocks.

Igneous rocks

Igneous rocks are the result of the cooling and solidification of molten material or **magma**, as it is termed, from deep below the earth's surface. Temperatures deep below the earth's crust are extremely high and rocks and minerals exist in a molten condition at depth. Sometimes magmatic materials are poured out on to the surface of the earth, as in the case of a volcanic eruption (see page 43), while under other conditions the magma fails to reach the surface, and cools and solidifies at depth. Igneous rocks may be distinguished from sedimentary and metamorphic rocks by their texture, structure, mineral content, and complete lack of fossils.

Rocks resulting from the cooling of magma on the earth's surface are termed **volcanic** or **extrusive** rocks. As the magma reaches the surface it loses its gaseous content and cools rapidly. This prevents the growth of large crystals, and a very fine-grained rock results. Examples include basalt, felsite, pumice and obsidian.

Rocks which formed from the cooling and solidification of magma beneath the earth's surface are known as **plutonic** or **intrusive** rocks. Because of their slower rate of cooling compared with volcanic rocks, they tended to develop a coarse texture, and are generally characterised by large mineral crystals. Examples include granite, gabbro and diorite. Granite is probably the most common of the intrusive igneous rocks and is composed of large crystals of quartz, felspar and mica.

Intrusive rock forms

Intrusive igneous rocks which have been injected into the surrounding rocks assume many different forms. Generally these are only evident after the overlying rocks have been removed by erosion.

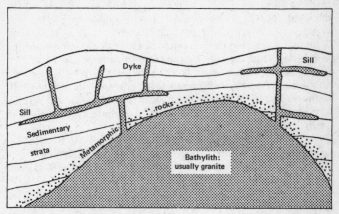

Figure 7. Types of igneous intrusion

Some of the most common forms of intrusion are described below:

Bathyliths or batholiths are enormous masses of granite which extend downwards to great depths in the earth. They are the largest type of igneous intrusion. When uncovered by the erosion of the overlying rocks they usually form plateaux with steep edges. Examples include the Wicklow Mountains (Eire), Dartmoor (Devon), Bodmin Moor (Cornwall), and the Cairngorm Mountains (Scotland).

Dykes are wall-like intrusions produced by magma rising upwards along vertical fissures which cut through existing strata. A well-known example is the Cleveland Dyke in Yorkshire. On the Isle of Arran dykes occur in 'swarms' and form minor headlands along the coast.

Sills are tabular bodies of igneous rock which have been intruded along the bedding planes of sedimentary rocks. A sheet of resistant igneous rock, parallel with the strata into which it has been intruded, is thus formed. Probably the best-known example is Whin Sill which runs across northern England and which was used for much of its length as the foundation for Hadrian's Wall.

Sedimentary rocks

All rocks exposed on the earth's surface are vulnerable to various agents of erosion. They may be attacked chemically, or may disintegrate as a result of mechanical processes such as the action of running water or frost. The rock fragments produced in these various ways are frequently picked up and transported by wind, water, or ice, and when the transporting agent has dropped them they are generally referred to as **sediment**. Sediment formed in this way may be deposited in the bed of the sea, in lakes, or on low ground. Over a long period of time it will build up, layer upon layer, and become consolidated by pressure and cemented together by chemical action to form sedimentary rocks. Rocks formed in this manner are termed **clastic** sedimentary rocks. Examples include clay, shale and sandstone.

In other cases sediments were precipitated from material dissolved in water to form **chemical** sedimentary rocks. Rock-salt, gypsum and anhydrite were all formed in this way.

Certain chemical sediments were deposited by or with the assistance of plants or animals, and these are termed **organic** or **biochemical** sediments. Carboniferous limestone, for example, is composed of the remains of minute marine organisms which extracted calcium carbonate from the sea water to build up their shells and skeletons. When these organisms died, they sank to the sea bed, their soft parts decayed, but their calcareous remains accumulated and eventually became compacted to form limestone. Chalk is thought to have been formed in a similar manner. Coal, which consists of the carbonised remains of plants, is another example of an organic sedimentary rock.

Sedimentary rocks can be distinguished from igneous and metamorphic rocks by the fact that they are generally deposited in layers or beds called **strata**. These layers are separated by **bedding planes**. Lines of weakness and fractures perpendicular to the bedding planes are termed **joints**.

Metamorphic rocks

Both igneous and sedimentary rocks may be changed into new forms as a result of being subjected to great heat and/or pressure. This process is called **metamorphism**, and the resultant rocks are known as metamorphic rocks. During the process of metamorphism the original rock undergoes physical, and sometimes chemical, changes which alter its texture, colour and structure.

Metamorphism may be initiated in various ways. If magma is

intruded into sedimentary rocks, the rocks in contact with the intrusion will be altered by the great heat to which they are subjected (see figure 7). This is referred to as **contact metamorphism**. Metamorphism also occurs when rocks are subjected to great pressure as a result of the mountain-building processes described in the following chapter. In this case complex and widespread changes take place. These are described by the terms **dynamic or regional metamorphism**.

The form of metamorphic rocks is determined by a number of factors; notably, the nature of the original rock, the type of metamorphism to which it has been subjected, and the intensity and duration of the metamorphism. Thus, some rocks may be only slightly altered, others radically changed. A great variety of metamorphic rocks are therefore found, and identification and classification is often difficult. Listed below are a number of common igneous and sedimentary rocks and their metamorphic equivalents.

Original rock	Metamorphic rock
Sandstone	Quartzite
Shale	Slate, schist
Limestone	Marble
Granite	Gneiss

Because of their mode of formation, metamorphic rocks are characterised by the obliteration of any former fossils or stratification. They are also much harder than the original rocks from which they were formed. In many cases minerals have been flattened and rearranged in roughly parallel bands which run through the rock. This is termed **foliation**.

Rocks and relief

The properties of the most common rocks should be learned, together with their distribution in the British Isles. The distinctive scenery associated with these rocks should have been studied, preferably in the field, or alternatively from OS maps and photographs. Examination questions on rock types most commonly relate to granite, carboniferous limestone and chalk. Study and learn sample areas for these rocks and be able to draw maps and diagrams to illustrate structure, relief, drainage, etc., as appropriate. Because of the complexity of metamorphic rocks, questions relating to them appear rarely, if ever. The following notes on four common rock types will provide an indication of the type of material which should be included in answers to questions on rocks and scenery.

Granite

Granite is a resistant igneous rock composed of quartz, felspar and mica. It usually forms areas of mountain or moorland. Dartmoor, for example, rises to a height of 622 m. It is an impervious rock, and granite areas are therefore characterised by numerous surface streams and rivers. The weathering of granite generally leads to the development of poor, thin, acid soils. Accumulations of peat may be found in poorly drained areas. Typical vegetation consists of rough grassland, peat bogs and scattered woodland. On Dartmoor, projections of bare rock on hill summits are known as **tors**. Examples include Hay Tor and Yes Tor.

Carboniferous limestone

The typically dry, upland landscape developed on areas of Carboniferous limestone is known as **karst** scenery. In Britain notable examples of upland limestone scenery are found in the Mendip Hills, the Peak District of Derbyshire, and the Malham district of Yorkshire. The distinctive scenery in such areas stems largely from two factors, the permeability of the rock and its solubility in rain and ground water. Joints and fissures become widened by solution and the surface often furrowed and fretted to form a limestone pavement with blocks and fissures known as **clints** and **grykes** respectively. Various holes and hollows known as **swallow-holes**, **swallets** or **sinks** are found at the surface. Some have streams plunging into them, such as Gaping Ghyll on the slopes of Ingleborough in Yorkshire, while others are completely dry. Streams diverted underground down swallow-holes often flow through **caves** which may contain **stalactites** and **stalagmites** formed by the deposition of calcium carbonate from percolating water. **Resurgent streams**, such as the River Aire which emerges from the foot of the limestone cliffs at Malham in Yorkshire, and **springs**, are commonly found at the base of limestone outcrops. **Dry valleys** are another typical landscape feature. Owing to the well-developed jointing in limestone, many such valleys are gorge-like in form. Some, including Cheddar Gorge in the Mendip Hills, have been attributed to the collapse of former cave systems. Soils in limestone areas are usually thin and alkaline, supporting poor grassland and few trees. Exposures of bare rock, which are common, are termed **scars**.

For most questions on limestone landscapes it is essential to show an understanding of the processes involved. Limestone contains a high proportion of calcium carbonate. The reaction between calcium carbonate and rainwater, which is a very dilute carbonic acid, produces the soluble calcium bicarbonate which is carried away in streams and ground water. Limestone is therefore a soluble rock, although the rate of solution is extremely slow.

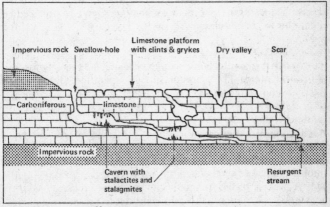

Figure 8. Features of karst scenery

Chalk

The scenic differences between areas of hard mountain limestone and areas composed of the much softer chalk arise principally from differences in porosity and permeability, which is very largely a product of jointing. Carboniferous limestone is permeable but not porous. The jointing in chalk, which is less well developed, exercises little control on relief, and the typical scenery consists of smooth, rounded, gently-rolling hills. Dry valleys are common features. Other valleys known as **winterbournes** are dry for most of the year, but contain surface streams after periods of heavy rain. The land use of chalk areas is very varied. Traditionally chalk areas are given over to sheep grazing, but in recent years there have been notable extensions of arable cultivation into many chalk upland areas. Soils are usually thin and dry. **Dew ponds** have been constructed in many areas to overcome the lack of surface water. Patches of **clay-with-flints**, a residual deposit left after the solution of the chalk, occur on hill summits and give deeper soils which often support a cover of beech woodland. Chalk areas in Britain include the North and South Downs, Salisbury Plain, the Chiltern Hills, Lincolnshire and Yorkshire Wolds, etc.

Millstone grit

Millstone grit is a dark-coloured, coarse, resistant sandstone. Millstone grit outcrops usually form areas of high moorland. Many of the hills have sharp ridges and steep sides often known as 'edges'. The typical vegetation in such areas consists of coarse grass, bracken, heather and cotton grass. Surface streams are abundant and in places peat moors and bogs are found. Reservoirs, giving soft water, have been built in many millstone grit areas. Typical gritstone scenery may be seen in the central Pennines west of Sheffield.

Key terms

Bathylith A gigantic dome-shaped intrusion of igneous rock.

Bedding plane The plane or surface separating the individual beds of sedimentary strata.

Dry valley A valley originally formed by stream action, but now without a surface stream.

Dyke A wall-like igneous intrusion caused by magma rising up through vertical fissures.

Igneous rock Rock formed by the cooling and solidification of molten material from beneath the earth's crust.

Joint A crack or fissure intersecting a mass of rock. In sedimentary rocks joints are roughly perpendicular to the bedding planes.

Karst The term is used to describe dry upland areas of limestone with distinctive surface and subterranean forms resulting from the action of water on the limestone.

Magma Molten material from which igneous rocks are formed.

Metamorphic rock Former igneous or sedimentary rock changed into new forms by being subjected to great heat and/or pressure.

Plutonic or intrusive rock Igneous rock formed beneath the earth's surface.

Sedimentary rock Rock formed by the compression and consolidation of mechanical, chemical or organic sediment.

Sill A tabular sheet of igneous rock injected along the bedding-planes of sedimentary rocks.

Stratum (plural: strata) A bed or layer of sedimentary rock.

Swallow-hole A surface opening, especially in limestone areas, through which a surface stream is diverted underground.

Tor An isolated mass of weathered rocks, usually granite, on the sides or summits of hills.

Volcanic or extrusive rock Igneous rock formed on the earth's surface.

Chapter 5
Earth Movements

Most of the rocks of the earth's crust are no longer found in the position in which they were first formed or deposited. Most have been subjected to enormous pressures known as **tectonic forces** which have deformed the original rock structures. The summit of Mount Everest, for example, is composed of limestone which must have been deposited and formed on the sea bed. As a result of the forces responsible for the formation of the Himalayas, rock strata must have been displaced and thrust upwards by many thousands of metres.

Tectonic movements usually occur slowly and imperceptibly over long periods of time, but some, as in the case of an earthquake, take place suddenly and violently. In some instances the rocks move vertically, causing uplift or subsidence of the land, while in other cases the rocks move horizontally or laterally as a result of compression or tension. Vertical displacements of rocks are referred to as epeirogenic movements, and lateral deformations as orogenic movements.

Epeirogenic movements are relatively slow movements involving the broad uplift or submergence of extensive areas. Normally the rock strata affected by such movements are not intensively folded or fractured, although they may be gently tilted or warped. The chief effects of such movements are best seen in coastal areas, where they produce changes in the height of the land relative to sea-level. Uplift produces raised beaches (locally) or wide coastal plains (regionally). Uplift also gives rivers new power to erode their valleys (rejuvenation). Subsidence of the land produces a coast indented with drowned valleys, known as rias, and estuaries.

Orogenic movements tend to be more intensive than epeirogenic movements and generally produce complex folding and fracturing of the rocks involved. These are the forces responsible for the creation of the world's major mountain belts such as the Andes, Rockies, Alps and Himalayas. Volcanic activity and earthquakes are closely associated with the orogenic type of crustal disturbance. The geological history of the earth has been characterised by a number of orogenic or mountain-building phases separated by long periods of relative stability. Thus, in Europe the various mountain regions are the result of the Caledonian, Hercynian and Alpine mountain-building phases.

Earthquakes

Earthquakes, which are physical convulsions of the earth's crust, provide indisputable evidence that crustal movements are still taking place at the present time. Some of these earth tremors are responsible for large-scale death and destruction, while others are so small that they can only be detected by means of an instrument known as a **seismograph**.

Earthquakes generally originate deep in the crustal rocks. The point on the earth's surface which lies immediately above the **focus** or point of origin is known as the **epicentre** of the earthquake. When an earthquake occurs, shock waves known as **seismic waves** spread out in all directions from the epicentre, decreasing in force and destructiveness with distance from the centre. Earthquakes, like volcanoes, tend to occur in a number of well-defined belts where the crustal rocks are unstable and subject to shifts and movements. It has been estimated that 80 per cent of the world's earthquakes originate in the mountain belts fringing the Pacific Ocean, and a further 15 per cent in a belt extending from the Mediterranean region to the Himalayas. Disastrous earthquakes causing widespread damage and great losses of life have occurred in recent years in Chile (1960), Skopje in Yugoslavia (1963), Alaska (1964), Peru (1970) and Turkey (1973 and 1975). Earthquakes which originate beneath the oceans often generate enormous seismic sea waves called **tsunami**. These waves, which may be up to 30 metres high, are capable of causing tremendous destruction in coastal areas.

Fold structures

Compressional forces acting along a horizontal axis in the earth's crust may produce a series of folds or wave-like structures in sedimentary rocks. Under surface conditions rocks are relatively brittle and will fracture when placed under stress. On the other hand, deeply buried rocks, subjected to high temperatures, are more 'plastic' and tend to warp or fold rather than fracture.

Fold structures vary greatly in complexity and size. Simple upfolds or arches are termed **anticlines**; downfolds are referred to as **synclines**. Many variations in the form and intensity of fold structures may be noted. Some of the more common forms are shown in figure 9. These include **simple folding** in which the strata have been compressed into a series of gentle anticlines and synclines, **asymmetric folding** owing to pressure being directed more strongly in one direction than another, **recumbent folding** resulting from exceptionally strong compressional forces, and **nappe structures** in which the folding of rocks has become so intense that fracturing and displacement have occurred along the axis of the fold.

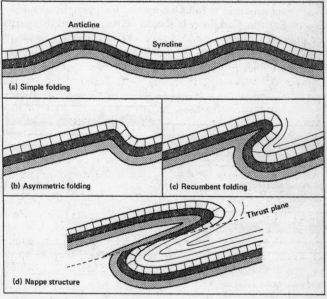

Figure 9. Fold structures

The formation of the various fold-mountain belts of the world is thought to have involved the accumulation of sediment in enormous downwarped troughs known as **geosynclines**. The sediment which thus accumulated over millions of years was later thrust upwards into a series of complex fold structures by compressional forces acting along a horizontal axis. The Alpine mountain-building forces compressed geological strata which were between 300 and 600 kilometres wide into a mountain chain 150 kilometres wide. Volcanic activity accompanied these mountain-building movements, and great quantities of igneous rock are found in the cores of fold mountain chains.

Fault structures

In some cases rocks respond to pressure by fracturing rather than folding, and moving along the lines of fracture. This is known as **faulting**. Simple fault structures involve movement and displacement along a single fracture. Two main types may be noted. A **normal fault** is produced by tensional forces whereby one block slips downwards relative to another, while a **reverse fault** is produced by compressional forces causing one block to override another (see figure 10). In both cases the vertical displacement of rock strata produces a **fault-scarp**, as, for example, along the Craven Fault in Yorkshire. It should be remembered that faults are lines of weakness, and, as such, are exploited by agents of erosion. For example, the long narrow valley of Glen More in north Scotland follows the line of a major fault.

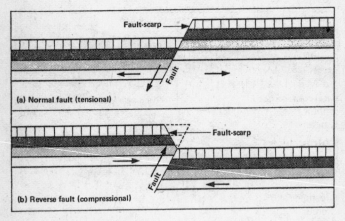

Figure 10. Simple faulting

More complex fault structures involving numerous faults may produce a variety of landforms. Where two faults run parallel a central block of land may be raised to form an upland area known as a **horst**. Examples include the Black Forest (Germany), the Vosges Mountains (France), and the island of Sardinia. A series of horsts in which some blocks have been uplifted more than others are referred to as **block mountains** (see figure 11). A good example is provided by the basin and range topography of the area between the Sierra Nevada and the Wasatch Range in the western USA.

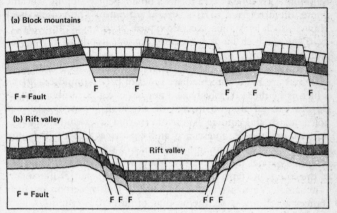

Figure 11. Block mountain and rift valley structures

In other instances a section of land may be lowered between two parallel sets of faults to produce what is termed a **rift valley** (see figure 11). Examples include the Rhine rift valley between Basle and Mainz, and the East African rift valley which extends from the Red Sea (itself fault-guided) to the coast of Mozambique, and contains many long, narrow lakes such as Lake Malawi, Lake Tanganyika, etc.

Volcanoes

As mentioned earlier, the processes of mountain-building are frequently accompanied by volcanic activity. Volcanoes are vents in the earth's crust through which molten rock (lava) and other volcanic products (gases, rock fragments, ash and dust, etc.) are ejected. If the lava reaches the surface through a single central vent it will normally form a volcanic cone. This type of volcanic eruption is termed a **crater eruption**. If, on the other hand, the lava is extruded through a long fissure it will simply spread out over the surrounding areas to form a lava plain or basalt plateau. This is termed a **fissure eruption**. Areas formed in this way include the Antrim plateau of Northern Ireland and the Columbia plateau of Oregon and Washington (USA).

Types of volcanic mountains

Volcanic cones or mountains vary in size and shape according to their age, the violence of earlier eruptions, and the type of lava being emitted. **Basic lava**, which has a low silica content, is very fluid and flows for a considerable distance from the point of eruption. This type of lava forms a broad, gently-sloping volcanic dome, often referred to as a **shield volcano**. The volcanoes of Hawaii, such as Mauna Loa, are typical of this type. In contrast, **acid lava**, which has a high silica content, tends to be viscous, and consequently forms a compact, steep-sided cone such as Mount Pelée on the island of Martinique. Between these two extremes are many intermediate types of cone, composed of lava which is neither excessively acid nor basic, e.g. Stromboli. **Composite cones** are constructed of alternate layers of lava and volcanic ash and cinders. These alternating layers indicate intermittent periods of quiescence and explosive activity. In some cases lava has burst through the flanks of the mountain to form subsidiary cones (see figure 12). Many well-known volcanoes are of the composite type, for example, Fujiyama (Japan), Kilimanjaro (Tanzania), Vesuvius and Etna (Italy). Under certain circumstances a volcanic cone, the product of numerous eruptions, may be virtually destroyed by a subsequent eruption of great violence. As a result, all that is left of the former cone is a roughly circular depression or crater known as a **caldera**. A well-known example is Crater Lake in Oregon, USA. On a much larger scale another example of a caldera is provided by the volcano of Krakatoa (between Java and Sumatra), which was destroyed in 1883 by one of the most violent and explosive volcanic eruptions in 'recent' times.

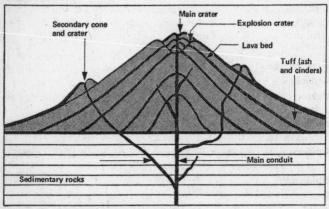

Figure 12. A composite volcanic cone

The distribution of volcanoes

Volcanoes may be described as active, dormant or extinct according to their relative state of activity. **Active** volcanoes are those which are continually or periodically in a state of eruption; **dormant** volcanoes are at present inactive but have been known to erupt within the last thousand years or so; **extinct** volcanoes are those which have not erupted within historic times. Large volcanic eruptions in recent years include those of Vesuvius (1943), Tristan da Cunha (1961), Surtsey (1963), Etna (1971) and Mauna Loa (1975).

The distribution of active and dormant volcanoes is very similar to that of earthquakes. The pattern should be studied and learnt by reference to an atlas map showing the distribution of seismic and volcanic features. A belt of volcanoes, sometimes referred to as the 'Fiery Ring of the Pacific', encircles the Pacific Ocean, and includes the volcanoes of South and Central America, Alaska, the Aleutian Islands, Japan, the Philippines, and the East Indies. Other regions of volcanic activity include the Mediterranean Basin, East Africa, the West Indies, Hawaii, Iceland and Antarctica.

Hot springs and geysers are found in many areas of present or recent volcanic activity. It has been suggested that **hot springs** are caused by ground water being heated by contact with rocks at very high temperatures lying at relatively shallow depth beneath the surface. Examples are found in Iceland, Yellowstone National Park (USA), and on North Island, New Zealand. **Geysers** are hot springs from which hot water and steam are discharged under great pressure, e.g. Old Faithful in the Yellowstone Park (USA).

Key terms

Anticline An upfolded or arch-like structure in folded rocks.

Block mountain A mountain or upland area consisting of uplifted, fault-bounded blocks.

Caldera A large depression formed by the destruction or subsidence of a former volcanic cone by later volcanic activity.

Epeirogenic movement A vertical movement of the earth's crust involving changes of level over a large area.

Epicentre A point on the earth's surface lying immediately above the focus or point of origin of an earthquake.

Fault A fracture in the earth's crust accompanied by displacement of rocks on one side of the fracture in relation to those on the other.

Fault-scarp A steep rock face resulting from the relative uplift of rocks on one side of a fault.

Geosyncline A large elongated trough or depression in the earth's crust in which sediments accumulated prior to being subjected to mountain-building forces.

Geyser A hot spring from which a column of hot water and steam is explosively discharged at intervals.

Horst An uplifted area bounded by faults.

Hot spring An eruption of hot water at the earth's surface. Common in areas of recent volcanic activity.

Nappe A mass of rocks thrust forward by a combination of intensive folding and low-angle faulting.

Orogenic movement A complex movement of the earth's crust involving the folding and fracturing of rocks. An earth movement of mountain-building scale.

Rift valley A valley formed by the sinking of land between two roughly parallel faults.

Seismic wave A shock-wave radiating out from the epicentre of an earthquake.

Shield volcano A large gently-sloping volcanic cone composed of fluid basic lava.

Syncline A downfolded or trough-like structure in folded rocks.

Tectonic Pertaining to movements of the earth's crust.

Tsunami Large sea waves generated by earthquakes.

Chapter 6
Weathering and Soils

Weathering

The term **weathering** refers to the disintegration and decomposition of rocks *in situ*. Care should be taken not to confuse the term with **erosion**, which implies the transportation of rock fragments and debris as well as the wearing away of the land surface by various agents such as rivers, glaciers, wind and sea. These latter processes are described in the following four chapters. Here we are concerned simply with the break-up of rocks on the spot and the related processes of soil formation. Weathering is a slow, unspectacular, but nevertheless important, geological process. It provides much of the material from which sedimentary rocks are formed; it is important in the shaping of surface land forms, and, as mentioned, is responsible for the formation of soil. Three types of weathering may be distinguished; namely, physical, chemical and biological weathering.

Physical weathering

Physical or **mechanical** weathering takes place when a rock is reduced to smaller fragments without undergoing any change in chemical composition. This may result from various mechanical processes.

Frost-shattering occurs when temperatures frequently move above and below 0°C. Water which may have collected in cracks and crevices in a rock alternatively freezes and thaws, expanding and contracting in the process, and creating pressures of up to 150 kilograms per square centimetre which, over a period of time, cause the rock to break up. This process is most common in high latitude and high mountain areas. In mountain areas the shattered rock fragments resulting from this process accumulate on slopes as **scree**. A similar process is known as **frost-heaving**. This occurs in unconsolidated rocks. Frost exerts pressures in an upward direction and may force rock fragments to the surface, as well as causing damage to building foundations, roads, etc.

Alternate heating and cooling of rocks, particularly in areas with a large diurnal range of temperature, is another important mechanical weathering process. Rocks in desert areas, for example, expand as they are heated in the daytime, and contract at night when they are frequently subjected to sub-zero temperatures. These processes, repeated over long periods of time, cause small cracks and crevices to develop in the rock, which in turn make it vulnerable to frost-shattering and chemical weathering. In the case of a rock such as granite the different minerals making up the rock expand and contract at different rates, resulting in stresses and pressures within the rock which cause it to break up. This process is known as **granular disintegration**. A more spectacular process is **exfoliation** or 'onion-skin weathering', as it is sometimes known. In this case thin flakes, or even curved slabs, of rock peel off from exposed rock surfaces. It is generally suggested that this is caused by the outer surfaces expanding and contracting by a greater amount than the main body of the rock, thereby causing stress and eventually fracture below the surface layer. Under certain conditions, exfoliation may continue until rocks are reduced to rounded, almost spherical, boulders. This is known as **spheroidal weathering**.

Chemical weathering

In contrast to physical weathering which simply produces smaller fragments of the original rock, chemical weathering produces rock materials which are basically different from the parent rock. The most common processes of chemical decomposition of rocks are oxidation, hydration, carbonation and solution.

Oxidation occurs as a result of certain rock minerals combining with oxygen in the air to form oxides. Rocks containing iron constituents are particularly susceptible to this type of weathering. The oxidation of iron compounds is responsible for the reddish-brown colour of many rocks exposed to the air. The process of oxidation results in the decomposition and break-up of rock surfaces.

Hydration is the term given to the process whereby certain rocks and minerals absorb water and react to produce new substances. For example, anhydrite may be converted to gypsum, and the felspar constituent of granite altered to clay minerals (including kaolin) by this process. As well as the chemical results of hydration, minerals also expand as they take up water, thus causing stresses within the rock which accelerate its physical breakdown.

Carbonation describes the process whereby carbon dioxide (CO_2), which is generally present in air, water and soils, unites with certain rock minerals and alters their composition as a result. The substances produced in this way (carbonates and bicarbonates) are relatively soluble and therefore easily removed by water.

Solution of various rocks and minerals is probably the most important single process of chemical weathering. Rain-water is a very dilute form of carbonic acid (H_2CO_3) formed by the solution of carbon dioxide from the air. As ground water its acidity is generally increased by humic acids formed by decaying vegetation and other organic material. The solution effects of these dilute acids are particularly important in the decomposition of limestones and dolomites. The carbonic acid converts calcium carbonate into calcium bicarbonate which is removed in solution to produce the characteristic hard water of limestone areas.

Biological weathering

The action of plant roots and burrowing animals also causes the disintegration of rocks. Tree roots, for example, penetrating rock crevices, can be as destructive as frost-shattering in breaking up rock structures. Plants also contribute indirectly to chemical weathering by the production of acids which in turn dissolve and attack many types of rocks. The activities of man might also be mentioned in the context of biological weathering. The large-scale disintegration of rocks involved in road construction, mining, quarrying, etc., could be included in this category.

Soils

Various agents of erosion (river, glaciers, wind, etc.) are constantly at work removing the layers of loose weathered rock which overlie the solid bedrock in any area. The bedrock which lies beneath the mantle of weathered material breaks up as it becomes exposed to the various agents of weathering. The rate at which the bedrock is weathered depends on its composition, structure and relief, and the climatic conditions to which it is subjected.

The most important product of the various processes of weathering is soil, which consists of decomposed and disintegrated bedrock which has been altered to such an extent that it can support plant life. Geology, climate and vegetation all influence soil development. Soils are obviously related to the parent material from which they were formed. Climate influences soils through the vegetation it controls, and through the moisture that is available for weathering and soil-forming processes.

Soil composition

Soil is not merely finely divided particles of rock. It is a complex mixture of **mineral particles** derived from the weathering of the parent rock, organic material known as **humus** derived from dead plants and animals, and contains a complex **fauna** including earthworms, bacteria, etc. It is permeated by **gases** and by **water** containing many substances in solution. Within the soil the various constituents react with each other, so that soil material is in a constant state of change and adjustment. Indeed, because of these changes certain writers have been led to describe the earth's soil cover as 'a living body'.

What is also important is the fact that the soil in any area exists in a very delicate state of adjustment. For example, if the surface vegetation is modified as a result of human interference, the amount and composition of the organic matter being returned to the soil alters; this in turn affects the soil fauna, and the soil, adjusting itself to the changed conditions, alters in almost every respect.

Description of soils

Soils, adjusted as they are to local conditions, vary enormously from area to area. Owing to their wide variety and complexity, description and classification is difficult. Many criteria have been employed as a basis for classification. These include the following:

Soil texture It is possible to classify soils on the basis of the size of the constituent particles. A commonly employed system involves carefully analysing the soil and describing it according to the proportion of sand, silt and clay particles. Material larger than 2 millimetres in diameter is classed as gravel or stones. Below this size are sands down to 0·02 millimetres, and silt down to 0·002 millimetres in diameter. Smaller still are the clay particles. According to the proportion of these particles soils may be described as sand, sandy-loam, silty-loam, clay-loam and clay.

Soil acidity Soils also vary in acidity. It is possible to indicate soil acidity by measuring the concentration of hydrogen ions which are provided by the acids in the soil. The symbol pH is used to indicate this concentration. Neutral soils, neither acidic nor basic, have a pH value of 7. Acid soils have low pH values, in some cases as low as pH3, while basic soils may have values as high as pH9.

Soil profile A soil profile is simply a vertical section taken through the soil from the surface to the parent rock. In very general terms the upper layer of the soil is referred to as **topsoil**, and the lower, more compact, layer as **subsoil**. In fact a more precise method of describing soil profiles has been developed. Each different kind of soil has a characteristic soil profile in which it is generally possible to distinguish a succession of **soil horizons** or layers. In a mature, well-developed soil it is usually possible to identify three horizons, which are commonly designated by the letters A, B and C. Sometimes these are subdivided.

The A-horizon, the uppermost layer in a soil profile, is the topsoil, which frequently contains large amounts of humus. Beneath it is the B-horizon or subsoil which in moist climates usually contains clay and iron oxides but relatively little organic material. The C-horizon, the lowest layer, consists of only slightly altered parent rock which merges into the unweathered bedrock. The development of distinctive soil profiles in major climatic and vegetation zones (see page 109) provides the basis for the description and classification of soils on a world basis.

The zonal concept of soil development

The zonal concept of soil development is based on the idea that regardless of the parent material (with a few exceptions such as limestone) soil types will develop according to the climatic conditions in which they exist. Students whose syllabus includes the study of soils should learn examples of the main types of zonal soils and be able to draw typical profiles for each. Four examples are described below.

Podsol soils These are particularly well developed in high latitude areas where long, cold winters check organic decomposition. Although the annual precipitation is generally low, the spring snow-melt releases water which causes iron and aluminium oxides to be washed down from the A-horizon (**leaching**), leaving it ashy-grey in colour. Minerals from the A-horizon are redeposited in the B-horizon which has a reddish-brown and/or yellow-brown colour and clay-like texture. Podsols are naturally infertile and require heavy applications of fertiliser to make them suitable for agriculture.

Brown forest soils These occur in middle-latitude areas with moderate rainfall and a natural cover of deciduous woodland. Leaching is less prevalent than in the case of podsol soils, so that the A-horizon, which is usually rich in humus, is normally brown in colour. However, some minerals are washed down to the B-horizon which may be darker than the lower part of the A-horizon. Soil fauna are plentiful and active.

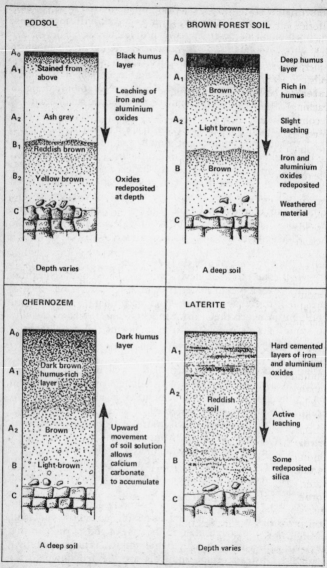

Figure 13. Profiles of podsol, brown forest, chernozem and lateritic soils

Chernozem soils (black earths) These are typical soils of the interior grasslands of the middle latitudes (prairies, steppes, etc.). They are characterised by a deep A-horizon, dark in colour and rich in humus. There is little leaching and the B-horizon is generally poorly developed. Chernozems are rich, naturally fertile soils.

Lateritic soils These are found in the humid tropics where leaching is particularly prevalent. The clay content of the soil decomposes and the silica constituents are washed down to the lower layers of the soil. Oxides of iron, aluminium, etc., remain in the upper layers. These silica-leached, reddish or red-brown soils are termed laterites. They are naturally infertile and very susceptible to soil erosion.

Intrazonal and azonal soils do not fit into the system of zonal soils. **Intrazonal soils** are not confined to any one zone. They develop where special factors such as poor drainage, or a parent rock with a dominant influence on soil development, prevents normal development. An example is the **rendzina** soil found in limestone areas. **Azonal soils** are young, immature soils which have had insufficient time to develop and adjust to climatic conditions. They include recent alluvial and marine deposits, glacial drift, recent volcanic deposits, etc. They generally lack a clear profile.

Key terms

Azonal soil A young, immature soil not adjusted to climatic conditions.

Erosion The destruction of rocks and the removal or transportation of debris by various agents such as rivers, glaciers, the sea, etc.

Exfoliation The splitting-off of the outer layers of rock surfaces. Especially common in desert areas.

Intrazonal soil A soil type which has developed under special conditions such as poor drainage, etc.

Leaching The removal of soluble soil constituents by percolating water.

Scree An accumulation of rocks or stones, often the result of frost-shattering, lying on hill slopes.

Soil profile A vertical section of soil showing the sequence of horizons from the surface to the parent rock.

Weathering The disintegration and decomposition of rocks at or near the earth's surface by mechanical or chemical processes.

Zonal soil A soil type related to the climatic conditions under which it has developed.

Chapter 7
Rivers, Valleys and Lakes

Running water is the most effective agent by which the surface of the earth is sculptured, except in deserts or frozen lands, where it rarely exists. Rivers and streams are channelled masses of water flowing over the land surface. They are nourished by direct precipitation, by surface run-off and by groundwater, which is simply precipitation stored in the ground and released during drier periods thus promoting a constant flow all year round. On average only about one-fifth of precipitation flows off in streams and rivers. Most of it is lost by evaporation.

The volume of water contained in any river or stream varies throughout the year according to rainfall and other climatic factors. Although many rivers flow continuously, in more arid countries smaller rivers may dry up seasonally. This occurs when precipitation is low, evaporation is high and there is an insufficient supply of water to maintain stream flow. At the other extreme, when the supply of water to a river exceeds the capacity of its channel, flooding occurs. When this is unexpected it may have catastrophic effects for the local population. On the other hand, the agricultural prosperity of much of Egypt depends on the regular and reliable flooding of the Nile.

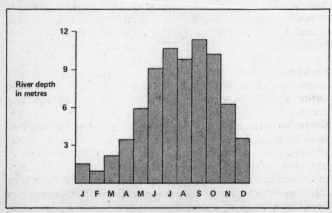

Figure 14. The régime of the River Yangtse at Ichang

The seasonal variation in the volume of water in a river is called its **régime**. River régimes have received much attention in recent years in connection with flood control and the production of hydro-electricity.

Fluvial processes

As rivers flow they perform the triple functions of transport, erosion and deposition. These are to some extent interdependent because the ability of a river to erode or deposit depends largely on the nature of the load it is transporting.

Transport

A river moves material in three main ways. Firstly, some soluble material is dissolved and carried in solution. Secondly, small particles are carried in suspension. And thirdly, rock fragments, stones and even boulders are rolled along the river bed, being transported both by gravity and the force of flowing water.

The energy of a river depends on its volume and velocity. The velocity, which is the speed at which a river flows, is determined both by its volume and by the gradient of its channel between source and mouth. As the velocity of a stream increases, for instance in time of flood, its capacity to move material increases enormously. This carrying capacity also depends on the nature of the load. A stream can normally carry a much larger load of fine material than of coarse, but in a few hours of flood a stream may move a greater weight of material than in several years of normal flow. In 1952 a flood of the River Lyn swept thousands of tonnes of boulders into the streets of Lynmouth in north Devon, an example of the suddenly increased transporting capacity of a fairly small river.

Erosion

A river erodes its bed by four interacting processes. **Hydraulic action** causes water to surge into cracks, sweeping out loose material and helping to weaken and break up the solid rock. **Corrasion** uses the load of the river for grinding and wearing away its bed and banks. This is especially effective where pebbles are whirled round by eddies in the hollows of the river bed, so cutting or drilling pot-holes. The general result of corrasion is to scour and excavate the bed. Furthermore, material derived by corrasion becomes part of the load and is used as a corrasive agent itself. **Attrition** consists of the wearing down of the load itself as the moving fragments constantly collide with each other. A progressive reduction in the size of transported material is the result. **Solution** is the solvent work of water as it flows over such rocks as limestone.

Deposition
There is a limit to the load which a river can transport. When a river attains a full load, vertical erosion of its bed must be balanced by deposition. At the same time lateral corrasion of the outside curve of a bank will be balanced by deposition on the inside curve in the slack of the current and, in this way, meanders begin to be formed. Deposition occurs when the rate of flow is checked, perhaps as the gradient lessens or where the river begins to curve. Material is deposited selectively, the largest fragments being dropped first and the smaller afterwards, so that there is a grading of material downstream.

The development of a river valley
As the processes of erosion and deposition continue, the overall shape of any river valley is constantly changing. This affects both the long-profile and the cross-profile of the valley. The actual form of a valley is largely determined by three factors: firstly, the erosive power of the river and its tributaries; secondly, the nature of the underlying rocks, their structure and their resistance to erosion; and thirdly, the stage which fluvial processes have reached in grading the long-profile of the river.

The long-profile of a river
The long-profile of a river is a section along its bed from source to mouth. Many rivers have their sources in the irregular relief of upland or mountain areas while their lower reaches are often on flat coastal plains. Consequently the gradient of a river bed tends to decrease from source to mouth although there may be irregularities in between where bands of resistant rock occur. So erosion tends to be more characteristic of the upper reaches of a river, and deposition of the lower. Generally a river's activity is devoted to attaining the smoothest possible slope and the most gentle gradient from source to mouth. Both erosion and deposition contribute to this goal. When it has been attained the river is said to have a **graded profile**. In reaches where erosion is active the river is said to be degrading its bed, while where deposition is occurring the river bed is being aggraded.

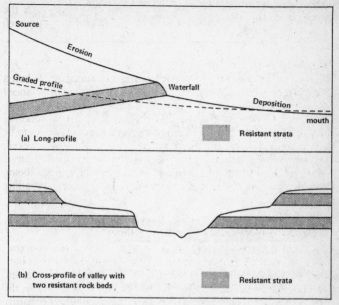

Figure 15. (a) Long-profile and (b) cross-profile of river valleys to show the effects of resistant strata on valley form

River valleys are sometimes considered to display characteristics of youth, maturity or old age according to the stage they have reached in their development. For many rivers these 'ages' apply to their upper or mountain, middle, and lower or flood-plain tracts.

Youthful stage

The source of a river is usually a **spring** occurring where the water table meets the surface of the ground. It may be near to the tops of hills in impermeable rocks or lower down in permeable rocks such as chalk or limestone. Rivers may also have their origins in lakes or glaciers. In a mountain course a fast-flowing stream usually cuts a steep-sided, V-shaped valley even though its volume may be small owing to its limited catchment area. If downward erosion is especially rapid, perhaps where a stream finds a structural weakness, a **gorge** may be formed. A young stream often follows a very winding course since it tends to flow round more resistant rocks or boulders.

These bends are gradually emphasised since the current is stronger on the outside of a bend and, as a result, **interlocking spurs** are formed.

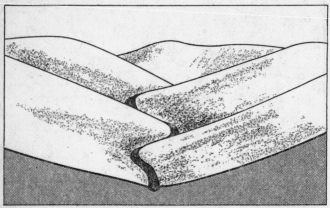

Figure 16. Interlocking spurs of a youthful valley

At this stage abundant rock waste is supplied to the river by weathering of the surrounding slopes. Active erosion of the river bed occurs especially by pot-hole formation and the long-profile is irregular with many rock ledges, rapids and small waterfalls.

Mature stage

In its middle course a river develops characteristics of maturity. Vertical erosion decreases and the river expends more energy on lateral erosion, widening its valley and wearing away the interlocking spurs which lie in its path. This widening is mainly the result of the erosive power of meandering which is particularly evident on the outer banks of bends. Meanders tend to grow and migrate downstream cutting into the bases of spurs and producing **river cliffs**. The long-profile becomes more even while the cross-profile shows a wider V-shape which is often asymmetrical. As spurs are worn away to **cusps** the area of the flood-plain is slowly increased producing a wider valley floor between steeper bluffs.

Old-age stage

An old river valley is characterised by a wide, flat **flood-plain** with a sluggish, meandering river. The long-profile shows only a very slight gradient. Large quantities of silt and gravel are

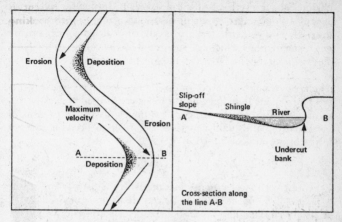

Figure 17. The development of meanders

deposited so that sometimes the river bed becomes choked and the river spreads out into several channels. Such rivers are described as **braided**. At the sides of a river channel, where the current is checked by friction with the banks, deposition of silt may be sufficient for embankments known as **levées** to be formed. Eventually deposition of silt in the river bed raises the

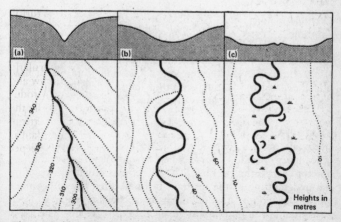

Figure 18. (a) youth, (b) maturity, and (c) old age in river valleys

channel above the level of the flood-plain. In flood a river may spread a thin veneer of silt or **alluvium** across the whole of its flood-plain. The great fertility of alluvium helps to explain the dense populations which are supported by the valleys of the Nile, Tigris-Euphrates, and many of the great rivers of Asia.

The meandering which began during maturity reaches exaggerated proportions in old age. Great loops are formed and, as lateral erosion of the outside of bends continues, these are eventually severed from the main course of the river to become **oxbow lakes** or **mortlakes**.

Deltas

When a river reaches its destination, which is usually the sea or a lake, much of its load is deposited as its rate of flow is suddenly checked and, as a result, a delta may be formed. This is most likely to occur where there is a good supply of silt, where salt water is present to bind particles together, and where offshore currents are weak. The deposits of silt at the mouth of the river gradually build up and an area of shallow water and mud-flats emerges. The main channel may become braided and split into several streams or **distributaries**. It is possible to distinguish various types of delta according to their shape. The most important are the **arcuate** type like the Nile, Po and Rhône, and the **bird's-foot or finger** type, of which the Mississippi is a good example.

Eventually lakes may be filled in by the growth of deltas. Crummock Water and Buttermere in the English Lake District were originally one lake and were divided in two by the formation of a delta. The largest lake deltas in the world are formed at the mouths of rivers like the Volga, Ural and Kura which flow into the Caspian Sea. But, except for the fact that the Caspian is enclosed by land, these are really marine deltas.

Waterfalls

Waterfalls are usually produced by the erosion of a river, mostly in its upper or middle course, where a band of more resistant rock outcrops across the river bed. The fall retreats slowly upstream by the undercutting of the resistant bed and often leaves behind a precipitous gorge. A good example is Niagara where a thick horizontal bed of hard limestone overlies softer shales. The fall is 52 metres high and the gorge 12 kilometres in length. The

Kaietur Falls in Guyana, some four times higher than Niagara, were formed owing to the Potaro River flowing over a ledge of resistant conglomerate overlying softer sandstones and shales. Many waterfalls are caused by igneous intrusions. High Force and Cauldron Snout are caused by the Whin Sill outcropping in the Tees valley.

Many waterfalls are found where there is a sharp, well-defined edge to a plateau. The Livingstone Falls in Zaire, in fact a series of thirty-two rapids, and the Aughrabies Falls on the Orange River, are examples of this in Africa. Waterfalls plunging from hanging valleys or over rock-steps are usually the result of glacial erosion.

Drainage patterns

Although no two drainage systems are exactly the same, for the sake of simplicity the basic patterns of rivers and their tributaries have been generalised into three main types, radial, dendritic, and trellis. **Radial drainage** occurs when streams flow outwards from a central upland mass. The drainage patterns of Dartmoor and the English Lake District are good examples. A tree-shaped **dendritic drainage** pattern develops on rocks of uniform resistance. On the other hand, where there are alternating outcrops of resistant and less resistant rocks a more angular **trellis drainage** develops.

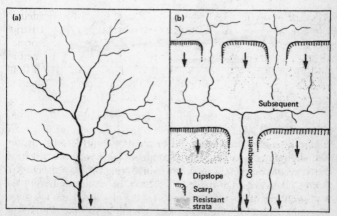

Figure 19. (a) Dendritic and (b) trellis drainage

Streams are sometimes distinguished according to their direction of flow and relationship with each other. Thus, the streams flowing down the dominant slope towards the mouth of the drainage system are called **consequent streams**. With a trellis system, tributaries flowing in conformity with the underlying rocks almost at right angles to the consequents are called **subsequent streams**.

Owing to differences in the resistance of rock to erosion, streams may expand their catchment areas by encroaching on other drainage basins or even by capturing other tributaries in the same system. For example, in scarplands, such as those of the Weald or the Paris Basin, where trellis patterns are common, a particularly active subsequent may cut back its valley and capture the headwaters of an adjacent consequent stream. This process is known as **river capture** and leaves traces in the form of **wind gaps** and right-angled bends in the courses of rivers.

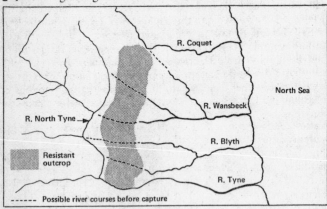

Figure 20. River capture in Northumberland

Lakes

A lake is a hollow on the earth's surface containing water. Some of the largest lakes, such as Superior and Victoria, are virtually inland seas and some, like the Caspian and the Dead Seas, are actually named as such. In arid countries lakes may dry up either partially or completely in summer or during prolonged drought. Some lakes are man-made, the result of damming rivers for hydro-electricity or water supply. Lake Volta in Ghana is a good example. Lakes are often classified according to how the hollow which they occupy was formed.

Lake-hollows produced by erosion

Some basins now occupied by lakes have been scooped out by the erosive power of valley-glaciers or ice-sheets. Many of the lakes in Snowdonia, the Lake District and the Lakes Plateau in Finland come into this category. Where soluble rocks like limestone are present, enough material may be carried away in solution to produce hollows in which lakes can form. The 'meres' of Cheshire are formed owing to the underground solution of salt deposits. Wind action may excavate hollows in deserts, like the Qattara Depression in Egypt, which later fill with water.

Lake-hollows produced by deposition

Lakes ponded up by deposition are known as 'barrier lakes' since they are usually formed as a result of a natural dam. Such a dam could be caused by an avalanche or a lava flow blocking a valley. Shallow lakes are often produced by deposition in delta areas. The Etang de Vaccares in the Rhône Delta is an example. Some types of depositional lakes result from glacial action. Terminal moraines have been instrumental in ponding up water, as in the case of Lake Garda in Italy. Thick deposits of boulder clay or glacial debris contain hollows in which small lakes may form. In addition, active glaciers may themselves act as dams and cause ice-barrier lakes in tributary valleys. There are many such lakes in Iceland.

Lake-hollows caused by earth movements and vulcanicity

The faulting, fracturing and warping of the earth's crust has produced hollows which now contain either salt- or fresh-water lakes. Many very large lakes, like the Caspian Sea or Lake Victoria, have this origin. Such lakes form in the narrow depressions of rift valleys; for example, the Dead Sea and Lake Malawi. Circular lakes sometimes occupy the craters of extinct or dormant volcanoes. Crater Lake in Oregon and Lake Toba in Sumatra are classic examples.

Key terms

Arcuate delta A delta with a rounded outer margin.

Bird's-foot delta A delta extended into the sea on either side of distributaries outstretched like fingers. The best-known example is the delta of Mississippi.

Braided channel A river course in which the water passes through a number of small, shallow, interconnected channels.

Consequent river A river whose course is directly related to the initial slope of the land surface.

Delta A roughly triangular-shaped area of low ground found at the mouth of many rivers, especially those discharging into lakes or seas with a small tidal range.

Dendritic drainage A pattern of branching streams and rivers forming a 'tree-like' plan.

Distributary A small river channel branching from the main river, as on deltas.

Flood-plain A plain, bordering a river, which has been formed from deposits of alluvium carried down by the river.

Graded profile A stream profile in which slope is delicately adjusted to provide just the velocity required for the transportation of the load supplied from the drainage basin.

Interlocking spur A spur of relatively high ground projecting into a river valley around which the river or stream meanders.

Levée A low bank or ridge built up by streams on their flood plains on either side of their channels. In some instances natural levées may be artificially raised to prevent flooding.

Oxbow lake A crescent-shaped lake caused by a river cutting through the neck of a meander loop during a period of flood. A common feature on flood-plains. Also referred to as cut-off lakes or mortlakes.

Radial drainage A system of streams and rivers diverging from a central upland area.

River capture The diversion of the upper waters of one river into another river system owing to the greater erosional power of the dominant river.

River régime The seasonal variation in the volume of water or discharge of a river.

Subsequent river A river whose course is related to an outcrop of relatively weak rock or a line of structural weakness. A tributary of a consequent river.

Trellis drainage A roughly rectangular pattern of drainage which develops in areas with rocks of varying resistance.

Wind gap A river gap or valley deserted by the stream which cut it, usually as a result of river capture.

Chapter 8
Glaciation

Today glacial processes affect relatively limited areas of the earth. Extensive ice-sheets exist in Greenland and Antarctica, and in lower latitudes smaller valley-glaciers are active in mountain ranges such as the Alps and the Himalayas, but in the past much larger areas of the earth were covered by ice. In Pleistocene times all of Britain north of a line extending roughly from London to Bristol was ice-covered and during the last stage of the Ice Age, which ended about 25,000 years ago, Scotland, Wales and northern England were ice-bound. It is obvious therefore that the landforms of large areas of northern Europe, North America and Asia have been affected by glacial action even though the permanent ice-cover has long since disappeared.

The formation of glaciers

On mountain slopes, above the **permanent snow-line**, snow will not completely disappear even in summer. Hollows, especially those shaded from the sun, will be permanently occupied by patches of snow. Melt-water from these patches penetrates the underlying rock and, by freeze-thaw action, enlarges the hollow. As the snow deepens the lower layers are compacted and turn, firstly, to **névé** and then, usually with the addition of frozen melt-water, into glacier-ice. When the frozen mass becomes large and heavy enough the influence of gravity causes it to move out of the hollow and downhill to form a **valley-glacier**. Where temperatures are low enough, as in Greenland and Antarctica, ice accumulation may occur on a much larger scale to form **ice-sheets**. These may be extremely thick. The ice at the South Pole is over two kilometres thick. Ice-sheets sometimes reach the coast and extend on to the sea, eventually breaking off and forming icebergs.

A series of valley-glaciers emerging on to a lowland plain may spread out and coalesce to form a single continuous ice-mass called a **piedmont glacier**, e.g. the Malaspina Glacier in Alaska.

Mountain glaciation

In some respects the work of glaciers is similar to that of rivers. Both are responsible for processes of erosion and deposition. Both are involved in the transfer of rock material from uplands to lowlands by which land-masses are slowly worn down. But this comparison cannot be taken too far since the landforms resulting from glacial erosion and deposition are quite different from those produced by fluvial processes.

Glacial erosion and transport

The hollow in which a valley-glacier originates is enlarged by freeze-thaw and by the slow movement of ice scraping away the sides and floor, plucking out boulders and transporting debris downwards. Eventually a large armchair-shaped hollow is excavated which may be alternatively called a **corrie**, **cwm** or **cirque**. Erosion is especially active on the back-wall of a corrie which may become very steep. Many corries are overdeepened below the outlet by which the glacier finds its way downhill. As a result in postglacial times many corries contain lakes. Blea Water Tarn in the Lake District is an example.

The material which is transported by a glacier in the form of rock fragments and gravel is called **moraine**. Quantities of moraine are carried on the surface of glaciers. This is supplied by freeze-thaw action on the valley sides above the glacier loosening fragments which fall on to the glacier forming **lateral moraines**. Where a glacier from a tributary valley coalesces with a valley-glacier two lateral moraines combine to form a **medial moraine**.

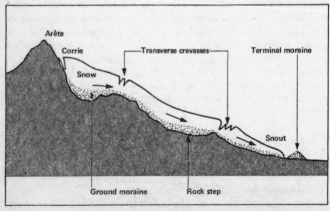

Figure 21. Section through a valley-glacier

66

The Aletsch Glacier in the Alps has six medial moraines, some of which are over 10 metres high. In addition to the moraines travelling on the surface of the glacier, **ground moraine**, mostly derived from erosion of the valley floor, is transported in the lower layers of ice. Eventually a glacier reaches a point at which the rate of melting balances the rate of downward movement. This lower extremity is called the snout of the glacier. Here all the morainic material which is being carried is deposited sooner or later and a **terminal moraine** is formed. Over a period of time the position of a glacier snout may change owing to variations in temperature or snowfall. As a result a series of minor terminal or **recessional moraines** are left corresponding with positions where the snout had been temporarily stationary.

The surface of a valley-glacier is often broken by **crevasses**. These cracks in the ice occur because of the different rates of movement of various parts of the glacier. **Transverse crevasses** occur particularly where there is a sudden change in the gradient of the valley floor.

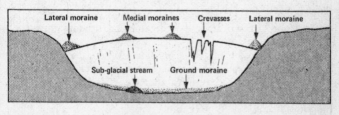

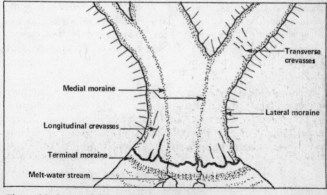

Figure 22. Cross-section and plan of a valley-glacier

Landforms caused by valley-glaciers

A mountain region which has been subjected to glacial erosion is recognisable by a number of distinctive landforms. On the rocky sides of steep mountains corries may have developed by the process already described. Ridges left between the back-walls of two corries are often sharpened into narrow edges called **arêtes**. A mountain peak may be eaten into by a number of corries eventually leaving a single **pyramidal peak** or **horn**. Snowdon and the Matterhorn are examples of this.

Glaciers which occupy pre-glacial river valleys greatly modify their form. The cross-profile tends to become **U-shaped**, with a flat floor and steep sides. A winding valley is straightened and any projecting spurs are planed off and **truncated**. Since their glaciers are smaller, downward erosion in tributary valleys is considerably less than in the main valley. Consequently, when the ice recedes, these are left as **hanging valleys** with streams that fall abruptly into the main valley in a series of cascades. Today this steep drop from tributary to main valley often provides a good site for a hydro-electric power station. The long-profile of a glacial valley sometimes shows a series of **rock-steps**, caused both by the unequal eroding power of different sections of the glacier and by differences in the resistance to erosion of the valley floor. Where the valley floor has been overdeepened, perhaps between two rock steps, a long, narrow **ribbon lake** may form after recession of the ice. Such lakes may also be caused by the deposition of terminal moraines. Lake Windermere and Coniston Water in the English Lake District are examples of ribbon lakes. **Fjords** may be produced by the drowning of glaciated valleys. A fjord often has an overdeepened floor – that of Sogne Fjord is over 1000 metres below sea-level – while a shallow threshold of solid rock or moraine is left at the seaward end.

A striking result of glacial erosion is the moulding of upstanding masses of rock. The upstream (or stoss) side of such a mass is smoothed and polished to a rounded shape while the downstream (or lee) side is made rougher by plucking. The resulting form is known as a **roche moutonnée**. A similar feature, known as **crag and tail**, is caused by a hard obstructive mass of rock. The crag lies in the path of oncoming ice and protects softer rocks in its lee from glacial erosion. An excellent example is the hard basalt plug of Edinburgh Castle Rock with its limestone 'tail'.

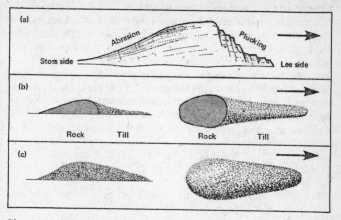

Figure 23. (a) Roche moutonnée, (b) crag and tail, and (c) drumlin features

Lowland glaciation

In mountain areas the main landforming role of glaciers tends to be erosive, while in lowlands glacial deposition is perhaps more apparent. Nevertheless, in some lowland areas the effects of erosion are widespread. This is especially true where an ice-sheet has moved vigorously, stripping away the soil and eroding irregularly to leave rock-basin hollows. The innumerable lakes of the Finnish Lakes Plateau, the Baltic Shield, and the Laurentian Shield in Canada are a result of this kind of action.

Landforms caused by glacial deposition

Just as a valley-glacier may deposit a terminal moraine across a valley, so a series of low hills extending for considerable distances over a lowland plain may represent the terminal moraine of an ice-sheet. Some terminal moraines extending across the North European Plain have been traced for hundreds of kilometres. Ground moraine deposited by a melting ice-sheet gives rise to flat or gently rolling topography such as characterises much of East Anglia. The material of such deposits is often called **boulder clay** or **till**. The composition of boulder clay varies enormously. Sometimes it is made up mainly of fragmentary material including large **erratic boulders** and sometimes mainly of finely-ground, greyish clay or sand.

Where a large mass of stagnant ice was left embedded in the boulder clay and later melted, a hollow known as a **kettle-hole** remains. This often contained a lake but most have been filled in by stream action since the Ice Age. Thick boulder clay was sometimes moulded into oval mounds called **drumlins**. These vary greatly in size, from a few metres to two kilometres or more in length, and in number, from isolated examples to large swarms such as exist in Northern Ireland and the Eden Valley. **Outwash sands and gravels** are deposited, not by the ice itself, but by the melt-water streams which issue from the snout of a glacier. A dwindling ice-sheet might lay down an extensive sheet of such sediments in the wake of its retreat. Another feature similarly formed has the appearance of a winding ridge of sand and gravel and is known as an **esker**. These may represent the deposits of streams flowing underneath the ice-sheet or, alternatively, the continuously receding 'deltas' of outwash material deposited at the ice edge by streams flowing on, in or under a rapidly decaying ice-sheet. They are exceedingly common in Finland, northern Poland and Sweden, where they wind across country among the lakes and marshes. They are also found in parts of northern England and Scotland.

Away from the ice-front **loess** deposits may be found. These consist of wind-blown dust derived from the glacial deposits in immediate post-glacial times. They exist in many parts of northern and central Europe and produce good soils for agriculture.

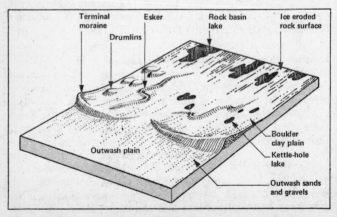

Figure 24. Features of a glaciated lowland

Glaciation and drainage diversion

The effects of glaciation are not limited to the erosional and depositional features outlined above. Over many of the areas which have been affected by glaciation, drainage patterns have been upset and rivers diverted to new courses. It is worth examining a few examples of such diversions in more detail.

Drainage diversion in North Yorkshire

During the Quaternary Ice Age the North York Moors were an 'island' surrounded by ice-sheets which extended down the east coast of England and through the Vale of York. In the vales between the protruding hills several lakes were ponded up. The lakes at the northern edge of the moors overflowed southwards cutting the striking overflow channel now known as Newton Dale. The earlier course of the River Derwent was eastwards to the sea north of Scarborough. However, when the coastal ice-sheet created a barrier which was later reinforced by the deposition of moraine at the eastern end of the Vale of Pickering, the result was a diversion of the River Derwent westwards to join the River Ouse near Goole.

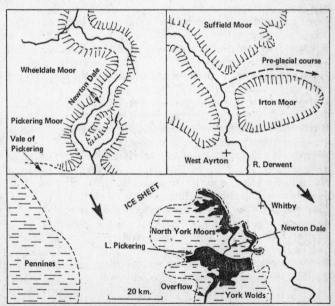

Figure 25. Drainage diversion in North Yorkshire

Diversion of the Upper Severn

In pre-glacial times the Upper Severn flowed to the estuary of the River Dee. This was blocked by ice moving from the north. A pro-glacial lake was ponded up which eventually overflowed through Ironbridge Gorge so that the Upper Severn was diverted southwards towards Gloucester and its present outlet into the Bristol Channel.

Key terms

Arête A sharp ridge or rock crest generally located between two corries.

Boulder clay Clay material containing rocks and stones deposited by former ice-sheet. Also known as till, drift, and ground moraine.

Corrie A more or less circular hollow or amphitheatre on a mountainside, sometimes containing a lake. Corries have steep back and side walls, but are open at their lower end from which a stream usually flows. Formed by glacial erosion. Also known as a cwm or cirque.

Drumlin A smooth, elongated hill or ridge composed of glacial deposits.

Erratic boulder A large boulder transported by ice from its original location.

Esker A long, narrow, winding ridge of sand or gravel deposited by former glacial streams.

Fjord A long, narrow, deep, steep-sided arm of the sea resulting from the flooding of a glaciated valley.

Hanging valley A tributary valley, the floor of which is higher than that of the main valley into which it leads.

Kettle-hole A surface hollow formed by the melting of former masses of ice embedded in glacial deposits.

Moraine An accumulation of rock debris deposited by glaciers or ice-sheets.

Névé Loose granular ice in the process of transition from snow to glacier ice. Also known as firn.

Piedmont glacier An extensive sheet of ice at the foot of a mountain range formed by the coalescing of valley-glaciers.

Ribbon lake A long, narrow lake occupying a glacially eroded rock basin. Also known as a finger lake.

Roche moutonnée A rock mass smoothed on one side by glacial erosion and left rugged on the other side.

Chapter 9
Deserts

Almost one-third of the land-surface of the earth experiences desert or semi-desert conditions, excluding the polar and sub-polar lands which are sometimes called 'cold deserts'. The hot or tropical deserts occur in the 'trade wind belt' between 20° and 30° north and south of the equator and tend to be on the western sides of continents. In addition, another group of arid lands comprises those areas of continental interiors which lie in the rain-shadow of high mountain ranges. These are sometimes called 'temperate deserts'; the Gobi desert and Patagonia fall into this category.

Deserts are characterised most obviously by aridity, normally receiving less than 250 millimetres of rain a year. This is reflected in the lack or scantiness of vegetation. In humid regions the work of the wind is limited because of close vegetational cover. In Britain, for example, it is confined to the movement of soil in the Fenlands and heaths of the Breckland of East Anglia and the movement of dry sand on the sea shore. In deserts, however, especially since the role of running water is limited, wind action is the major agent of erosion, deposition and mass-movement.

Wind erosion and transportation

Mechanical weathering is especially active in deserts because of the great diurnal range of temperature. It causes the initial disintegration of rock to form a mantle of rock waste which is susceptible to transportation by the wind. Wind erosion involves the triple processes of deflation, abrasion and attrition working together.

Deflation

This is simply the blowing away of dry, unconsolidated material, which gradually lowers the land surface. The finest material, borne high in the air, may be carried great distances. Saharan dust sometimes falls in France. Sand grains are swept along in 'sand storms' while still coarser material moves along in a series of 'hops' near the surface. One of the results of such action is the scouring out of **deflation hollows**. The floor of the Qattara Depression in Egypt has been excavated to over 140 metres below sea-level. The Kalahari and West Australian deserts have similar, though shallower, depressions.

Abrasion

The sand-blast effect of the wind carrying a load of hard sand grains is of great erosive power. It can smooth and polish a rock surface where resistance is uniform, or, where rock is of varying hardness, produce grooving, fluting and honeycombing. Heavier particles are carried near the ground but their force is retarded at ground-level by friction, so that undercutting is most marked about one metre above the ground and may leave **pedestal rocks**. In Britain some Pennine outcrops, such as the Brimham Rocks near Harrogate, show similar undercutting by the wind. Where a hard horizontal stratum lies above a soft one and erosion along joints penetrates through the hard cap, wind abrasion will carry on until separate tabular masses are left standing upon the softer underlying rock. These masses are known as **zeugen**. Alternatively, when rocks of differing resistance occur in roughly parallel bands, wind erosion can carve out gullies in the softer rocks leaving long ridges in between. The fantastically shaped **yardangs** of Central Asia have been formed by this kind of erosion.

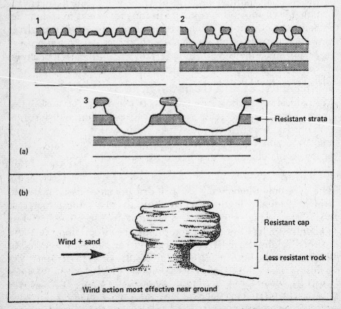

Figure 26. Formation of (a) zeugen and (b) pedestal rocks

Over a long period of time wind abrasion and deflation may continue to such an extent that only a few isolated, residual masses of rock are left standing above the general surface. Such hills are called **inselbergs**. In the deserts of western USA they are known as **mesas** or **buttes**.

Attrition

This is the process by which wind-borne fragments constantly colliding with each other gradually grind themselves down into rounded grains of sand. This sand forms the main end-product of wind erosion and the constituent material of the extensive sand deserts.

Deposition

Many factors account for the variety of depositional landforms found within desert areas. The nature of the surface over which the material is moved is important – whether it consists of deep sand or of bare rock, and whether there are surface obstacles such as bushes or rocks to arrest the movement of wind-blown particles. Even a dead animal may provide the nucleus for the formation of a sand-dune. The presence or absence of vegetation is very important. Trees are often deliberately planted to protect oases from incursions of moving sand. Even in Britain the planting of trees and shrubs is used as a method of stabilising coastal sand-dunes. The direction and strength of prevailing winds is another major factor. Dunes are normally aligned according to wind direction.

Dunes

The accumulation of sand into low hills of varying shape and extent is one of the most interesting features of desert relief. The simplest form is an **attached dune** which is caused by an obstacle such as a rock in the path of the wind. As the build-up of sand interrupts both the air-flow and the passage of sand, other smaller dunes may be aligned with the original. Perhaps the most distinctive sand-dune found in deserts is the crescentic **barchan**. These dunes are aligned transversely to the wind. Their 'horns' are trailed out in the direction in which the wind is blowing because there is less mass to be moved at the edges and the rate of movement is greater there. They may achieve a height of 30 metres and a length of more than 300 metres. Normally the slope is gentle on the windward side and steeper on the sheltered 'slip face' or leeward side which is slightly concave owing to eddying. The barchan will advance when the wind is strong enough to move sand up the gentle windward side and over the crest.

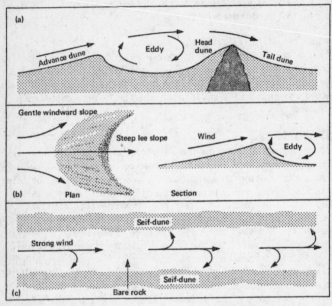

Figure 27. Formation of (a) attached dune, (b) barchan and (c) seif-dune

Unlike the barchan, some sand-dunes are aligned with the direction of prevailing winds. Such dunes usually occur in the form of a long ridge, often many kilometres in length, and are known as **seif-dunes**.

Loess

The finest wind-borne material may be laid down far beyond the limits of deserts. Deposits of such material, called loess or sometimes **limon**, were first studied in north-west China where they cover an area larger than France. In places these deposits are more than 300 metres thick. The loess sheet of north-west China was formed, over a long period of time, from fine material carried by outblowing winds from the Gobi desert. Increased rainfall as it moved seaward may have helped to wash it down from the air to the earth, where the steppe vegetation helped to stabilise it.

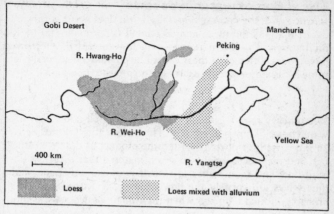

Figure 28. The loess area of northern China

The name 'loess' is applied in Europe to deposits of wind-blown dust derived, towards the end of the Ice Age, from the unconsolidated sands and clays laid down by ice-sheets. It is found in parts of Germany, Belgium and northern France but is absent from the most recently glaciated areas. Thick deposits of similar material also occur in North America, especially in the river valleys of the Mississippi and Missouri. Again, it is absent from Canada and the recently glaciated areas.

Desert landscapes

The interplay of the factors and processes already discussed leads to a great variety of desert landscapes. These, however, may be grouped into four main types.

The first is the true sand desert known as **erg** in the Sahara and as **koum** in Turkestan. Characteristically it consists of vast, almost horizontal, sand-sheets or of regular lines of dunes. The second is the stony desert, known as **reg** in the Sahara, where horizontal sheets of pebbles and angular gravel cover the surface. The third type is the rock desert, or **hammada**, which consists of a bare rock surface swept clear of sand and stones. The fourth type is the arid mountain range such as the Tibesti Mountains in the Sahara, the mountains of western Arabia and the peaks of Sinai. Here the terrain is extremely rugged with harsh, serrated outlines and steep, craggy faces cut into by **wadis**.

The action of water in deserts

In true hot deserts rain seldom falls, but it does occur at times. In desert margins the mean annual rainfall is normally 300 to 400 millimetres. This often represents a few torrential, short-lived downpours. Often even these cannot be guaranteed annually, and some areas may experience virtually no precipitation for a number of years.

When heavy rain does fall it may be channelled into steep-sided watercourses known as **wadis**. The torrent thus formed carries an immense load of solid matter, the product of desert weathering, and may sweep along so much material that it turns into a mud-flow and soon comes to rest. It is unlikely that the wadis themselves are entirely the product of such action. They were probably formed long ago when the climate of deserts was more humid than it is now.

Where a valley opens out on to lower ground an **alluvial fan** is deposited and a number of these may coalesce to produce a **bahada**. Large basins of inland drainage are a feature of almost every desert. They are often characterised by mud-sheets, which after rain may accommodate a temporary salt-lake and are normally surrounded by sheets of rock-salt or gypsum. These basins are known as **bolsons** in Mexico, **playas** and **salinas** in the United States and **shotts** in North Africa. The names are applied to both the temporary lakes and the basins which they occupy.

Another feature of arid and semi-arid areas is a gently-sloping surface of rock, bare or with a thin veneer of debris, that stretches away from the foot of an upland mass. They are superficially similar to bahadas but they are produced by erosion rather than deposition. They are known as **pediments**. Their origin is still not wholly understood but they may have been formed by the lateral erosion of heavily-laden streams emerging from mountain ranges into desert basins, continually changing course as their distributaries were blocked.

In addition to the role of rainfall, the presence of dew contributes much to weathering of a chemical type. The great diurnal range of temperature experienced in desert lands facilitates the formation of dew.

Key terms

Alluvial fan The alluvial deposit of a stream where it issues from a gorge on to open ground.

Bahada A belt of alluvial deposits found at the foot of high ground and produced by the coalescing of a number of alluvial fans. Also spelt 'bajada'.

Barchan An isolated crescent-shaped sand-dune with the horns of the crescent projecting downwind. Common in the deserts of Turkestan.

Deflation The removal and transportation of soil and rock particles by the wind.

Deflation hollow A shallow depression or hollow produced by the process of deflation.

Erg A sand desert, especially of the Sahara. Similar deserts are referred to by the term 'koum' in Turkestan.

Hammada A rocky desert with a surface of bedrock smoothed by abrasion and devoid of surface deposits as a result of deflation.

Inselberg An isolated residual hill common in many arid areas.

Loess A fine, loamy soil usually deposited by the wind. Also referred to as limon in some areas.

Pedestal rock A residual or erosional rock mass supported by a relatively slender neck or pedestal. Produced by wind erosion.

Playa A shallow temporary lake found in desert areas. Such lake hollows are dry for most of the year and often contain deposits of salt or gypsum. Also known as bolsons, salinas and shotts.

Reg A stony desert with a surface of small stones and gravel exposed by the blowing away of fine sand.

Seif-dune A linear or longitudinal dune found in desert areas where the wind blows consistently in one direction.

Wadi A steep-sided valley or gorge found in arid areas which is occasionally occupied by a stream or river.

Yardang Rock masses of irregular form in which grooves or depressions have been cut by wind erosion.

Zeugen Tabular rock masses in which the softer strata have been undercut and eroded by the wind.

Chapter 10
Coasts

Waves, tides and currents

Coastlines are undergoing constant changes owing to erosional and depositional processes brought about by the action of waves, tides and currents. Although tides have little direct effect on the coastline they are of considerable importance since tidal movements cause the action of waves to extend over the whole area between high- and low-water marks – the inter-tidal zone. Currents rarely cause any significant erosion but are instrumental in the movement and deposition of sand, pebbles and other material along the shore. By far the most important agent of coastal change is wave action, which can be either destructive or constructive depending on the weather conditions. On stormy days the power of waves can be very great indeed, and they are capable of breaking down cliffs, sea-walls and buildings in a short period of time. However, when the weather conditions are calm, smaller waves can act to build up material on the shore to form beaches and the other features of deposition.

Wave formation

Waves originate in the open sea, where the action of winds passing over the surface of the water sets up a series of undulations and then circulations in the body of the water (see figure 29). Wave length (the distance between one wave-crest and the next) depends upon the speed and duration of the wind, but wave height (the vertical distance between crest and trough) increases with the distance over which the wave has travelled (**fetch**). Thus, the potential height of waves on the Atlantic coast of Britain is much greater than on the North Sea or Channel coasts, since the waves may have travelled a much greater distance.

On reaching the shore the base of the waves comes into contact with the beach, causing friction and a reduction in speed at the wave base which eventually results in the crest breaking forward on to the beach. This rush of water up the beach is called the **swash**, and the retreat of water back down the beach which follows it is known as the **backwash**. Friction with the beach has another important effect, since unless the approach of waves to the shore is exactly parallel to it, friction will cause a refraction in the alignment of the waves towards the alignment of the shore.

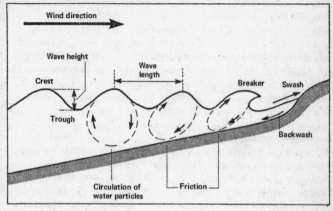

Figure 29. Wave motion

Erosional features

Processes of erosion

There are three major processes of coastal erosion. Firstly, the sheer pressure of breaking waves (over 29,000 kilograms per metre^2 in Atlantic storms) acts directly on the shore, and also indirectly by compressing air trapped inside rock joints and crevices which then expands with great force as the waves retreat. Secondly, rock fragments in the water are hurled against the base of cliffs by the waves, wearing these back by the process of corrasion. The debris thus produced is then broken down further by attrition, the grinding of particles against each other, to produce the typical beach materials of sand and shingle. The third major process is one which acts only on coasts where soluble rocks are exposed along the shore. For example, in the case of limestone, erosion results from the solution by sea water of calcium carbonate in the rock, causing it to weaken and disintegrate.

Cliffs and wave-cut platforms

Any steep rock face exposed on the shore may be termed a **cliff**, and this normally occurs where a ridge of high ground meets the sea, as at Beachy Head on the English Channel coast. The form of the cliff face and the rate of erosion is determined by the angle of dip, jointing and resistance of the rocks. Where the beds dip seawards erosion will be more rapid and cliffs less steep than on coasts where the beds dip landwards, since in the former case the rocks will tend to slip seawards along the bedding planes. The erosional action of the sea occurs within a relatively small zone at the base of the cliffs, where a notch is cut to undermine the cliff face, thus causing it to become unstable and collapse. As the cliff face recedes it leaves behind an eroded base, the **wave-cut platform**, on which beach material may be deposited, and which reduces wave action on the cliffs by causing waves to break earlier, thus slowing down the rate of cliff retreat.

Caves, blow-holes and geos

The erosive action of the sea picks out and excavates zones of weakness along the cliff base, often related to fault lines or jointing, and this may result in the formation of **caves**, such as those at Flamborough Head (Yorkshire) and on the Pembroke-shire coast. Pressure of water may then cause the formation of a shaft from the top of the cave to the land surface, a feature known as a **blow-hole** or **gloup**. Continued erosion results in the collapse of the cave roof to form a long, narrow inlet along the line of weakness, a feature known as a **geo**.

Arches, stacks and stumps

Where caves are formed on both sides of a headland, continued erosion may cause them to unite before their roofs collapse, thus forming a **natural arch**. A fine British example is Durdle Door on the Dorset coast near Lulworth. The collapse of an arch leaves an isolated pillar of rock standing away from the cliff face, a feature known as a **stack**, the most spectacular British example being the famous Old Man of Hoy in the Orkneys. Erosion at the base of stacks continues, which causes their eventual collapse, leaving behind residual features known as **stumps** below high-water level. The Needles, a large group of rocks off the Isle of Wight, illustrate the gradation in size from stacks to stumps, in this case formed out of chalk.

Headlands and bays

Whilst the lines and zones of weakness within a single rock type influence the development of small-scale coastal features, the general form of a coastline may be determined by the differential erosion of varying rock types. This effect is most pronounced where relatively resistant rocks, such as limestone, alternate with weaker rocks such as clay. The softer beds are eroded to form

82

bays and inlets, with the others remaining as **headlands** or capes. Swanage Bay in Dorset, lying between Ballard Point and Peveril Point, is one of a number of examples along the south coast of England.

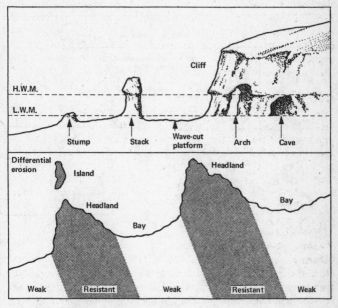

Figure 30. Features of coastal erosion

Depositional features

Beaches

Material which has been removed from the land by the processes of erosion may be transported along the shore to form features of coastal deposition, the commonest of which is the **beach**. Typically these are built up in relatively sheltered positions such as bay heads, but beaches are also to be found on more exposed coasts. One of the main processes by which beach material is moved along the coast is that of **longshore drifting**, which is caused by the oblique movement of waves on to the shore. The swash moves eroded debris obliquely up the beach whilst the backwash acts at right angles to the shore, hence there is a net movement of material in the direction of the dominant wind.

Wave action also results in the sorting of beach material according to size, with the coarser particles being deposited at the top of the beach whilst sand and silt are carried and deposited further down the shore by the backwash.

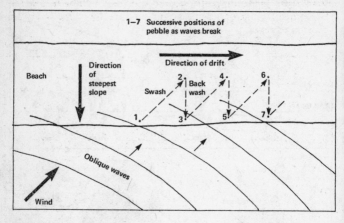

Figure 31. Longshore drift

Dunes

When the force of drying onshore winds acts on a large area of sand exposed on the beach, particles of sand may be carried inland by the wind and deposited to form coastal dunes. In some areas, such as Les Landes in south-west France, dune belts cover large areas and reach considerable heights, and unless steps are taken to stabilise the sand the onshore winds may cause the dunes to advance inland and engulf farmland and even villages and towns. Methods of stabilisation include the planting of coarse, sand-binding plants such as marram grass, or the establishment of tree plantations – notably pines.

Spits and bars

Where there is an indentation formed by a bay or estuary, or where the coast changes direction sharply, the action of longshore drift often causes beach material to be deposited and to build out beyond the original alignment. The feature thus formed, attached to the land at one end and projecting into the sea at the other, is called a **spit**. Spurn Head, at the mouth of the Humber, and Orford Ness in Suffolk, are examples, their form constantly changing as more sand is added and some parts eroded. Sometimes a ridge of sand or shingle is deposited right across a bay or estuary to form a **bar**, as at Slapton Ley in Devon. Chesil Beach, Dorset, is a long beach which has extended to link the mainland to the Isle of Portland, and features of this type are called **tombolos**. On some gently sloping shores an **offshore bar** may be formed, a ridge of beach material not directly connected to the coast at either end.

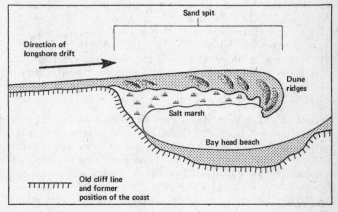

Figure 32. A sand spit

Salt marshes

One consequence of the formation of spits and bars is that the water on their landward side becomes sheltered from the direct action of the sea. Under such conditions fine sand and silt is deposited in the calm water, and this deposition gives rise to a favourable environment for colonisation by vegetation. The plants themselves then assist in encouraging further deposition by trapping particles, and thus the process of salt marsh formation is brought about. Examples of this can be seen on the landward side of Scolt Head and Blakeney Point in north Norfolk, though some of the marsh has been artificially reclaimed by drainage.

Changes in sea-level

Sinking or uplift of the land and rise or fall in sea-level have an important effect on the nature of coastlines. A classification can be made on this basis by distinguishing coasts of submergence and coasts of emergence.

Coasts of submergence

The sinking of the land-surface or a rise in the sea-level produces several distinctive features caused by the flooding of valleys and valley systems, the form of which depends on the nature of those valleys. The simplest form is that of an **estuary**, caused by the submergence of a major valley. Where a whole dendritic drainage system is flooded a complex series of inlets is formed, taking on the pattern of the valleys, and this feature is called a **ria**. Rias typically occur where an upland coast has been submerged, such as Plymouth Sound and numerous other inlets in Devon and Cornwall. Where a glaciated coast has been submerged the inlets take on the characteristics of glaciated valleys: long, deep and with steep sides. Such inlets are called **fjords**, and at their seaward end a ridge of rock forms a shallow section, the threshold, whilst off the coast are numerous small islands or skerries. Examples of fjord coasts are those of Norway and New Zealand (South Island).

A distinction can be made between those coasts where the main trend of the rock beds is parallel to the coast and those where it is at right angles. The former type is termed a **Dalmatian** or Pacific coastline and is characterised by a series of T-shaped inlets and elongated offshore islands, formed as the sea breaks through each resistant band of rocks and erodes the soft beds behind, as at Lulworth Cove in Dorset. The other type is known as an Atlantic coastline and is characterised by long inlets and headlands in alternation, as along the south-west coast of Ireland.

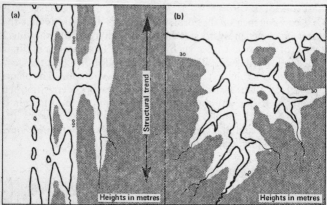

Figure 33. Types of submerged coastline: (a) Dalmatian coast and (b) ria coast

86

Coasts of emergence

On a lowland coast a fall in sea-level or an uplift of the land surface produces a wide, smooth, gently-sloping **coastal plain**, such as that of the south-eastern USA. The emergent deposits often form extensive beaches and dune belts, whilst mud flats and salt marsh may form on the shallow offshore zone. On upland coasts of emergence pronounced features of coastal erosion may survive inland as relict landforms. Former wave-cut platforms become **raised beaches** above the present beach level, and may have clearly identifiable cliffs, caves, stacks, beach deposits and so on. However, these tend to survive as relatively local features, rather than continuously over long distances, and are subject to normal weathering processes. In Britain the best examples are on the west coast of Scotland, where raised beaches at different levels can be correlated with particular periods of uplift.

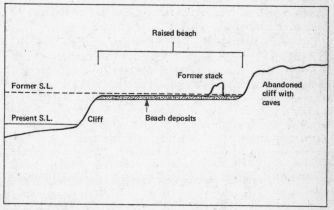

Figure 34. A raised beach

Key terms

Bar A low ridge of sand extending across an embayment of the coast. An offshore bar is a low sandy ridge running parallel to the coast but not connected to it.

Blow-hole A small opening on a cliff through which air and water are forced by the rising tide.

Dalmatian coast A coast running roughly parallel to the main structural or fold lines, and characterised by long, narrow offshore islands and T-shaped inlets.

Fetch The distance that waves travel before they reach a coastline.

Fjord A long, narrow, deep, steep-sided arm of the sea resulting from the flooding of a glaciated valley.

Geo A narrow coastal inlet usually related to joints, faults, dykes or other lines of weakness.

Longshore drifting The movement of sand and shingle along a beach by the action of oblique waves.

Raised beach A former beach or wave-cut platform now situated above the present sea-level, generally as a result of the land being elevated.

Ria A long winding inlet of the sea with numerous branching channels. Formed by the drowning of a river valley.

Salt marsh An area of low marshy ground periodically flooded by the sea and colonised by salt-loving plants.

Spit A low area of land projecting into the sea, joined to the mainland at one end and terminating in open water. Usually the result of longshore drifting.

Stack A high column of rock rising out of the sea and detached from the mainland.

Tombolo An island linked to the mainland by a beach or spit.

Wave-cut platform A gently-sloping rock surface cut by wave action.

Chapter 11
The Oceans

Salt water covers about 70 per cent of the surface of the earth. The Pacific Ocean alone has an area eight times greater than that of the USSR. Man has long been interested in the oceans both as a source of food and as a highway for transportation. The geographer is concerned with a number of aspects of the oceans and seas. These include the relief of sea and ocean beds, the accumulation of sediments upon them, changes in sea-level which may affect navigation, and the various movements of sea-water in the form of waves, tides and currents which influence adjacent coastlands climatically and in other ways. More recently the potential of the sea-bed for yielding valuable minerals such as oil has been realised. The economic role of the oceans with regard to food-supply is also increasing.

Submarine relief
The relief of the ocean floor may be divided into four main elements: the continental shelf, the continental slope, the deep-sea plain, and the deeps.

The continental shelf
Adjacent to the coasts of most continents between low-tide level and the 100-fathom mark is a shallow platform known as the continental shelf. Its angle of incline seawards is normally less than one degree. It is well developed off Western Europe, extending westwards for 250 kilometres from Land's End. Off the Arctic coast of Siberia the continental shelf is over 1,000 kilometres wide. The valleys of many rivers appear to continue across the continental shelf, suggesting that it was once part of the adjacent land-mass and was later submerged.

The continental slope
At the edge of the shelf the seaward slope steepens considerably, forming the continental slope, which descends to about 2,000 fathoms. The slope-angle here usually varies between two and five degrees but occasionally there are much steeper gradients.

The deep-sea plain
Almost two-thirds of the entire ocean floor lies at depths of between 2,000 and 3,000 fathoms. The relief of this area varies

considerably from a smoothly undulating plain to more rugged ridges and plateaux with an occasional volcanic peak reaching towards or even breaking the surface as an isolated island.

The deeps

The ocean floor is also characterised by **troughs** which descend to even greater depths. Although the word 'trench' is frequently applied to them, it is rather misleading since their sides slope fairly gently at inclines rarely exceeding seven degrees. Some of the deeps of the Pacific reach 5,000 fathoms below the surface. In terms of area the deeps occupy only a very small part of the ocean floor.

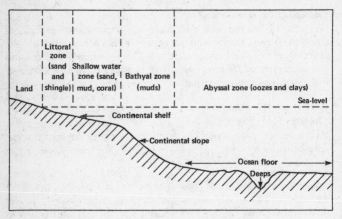

Figure 35. Configuration and composition of the ocean floor

The oceans

There are four oceans: the Pacific, the Atlantic, the Indian and the Arctic. A fifth, the Southern Ocean, is sometimes individually defined as the waters south of latitude 40°S.

The Pacific Ocean

A continental shelf is either absent from or very narrow in the Pacific and its bordering seas. Only on the western side, where the Seas of Okhotsk and Japan and the Yellow Sea have relatively shallow water, is it extensive. Depths increase very rapidly away from the western coasts of the Americas. In fact, the Pacific occupies an immense depression in the earth's surface. The origin of this depression has long been a matter for debate, and various theories have been put forward. One such theory holds that the Pacific hollow remained when the moon was torn away from the earth. Another postulates that the Pacific was occupied by a land-mass which slowly sank. It is difficult to test these theories, however, because little is known about the rocks of the ocean floor, and the great depth of oozes which cover them obstructs exploration.

In the western Pacific a series of submarine ridges breaks the surface to form a number of island arcs. These are often juxtaposed to long, narrow ocean deeps such as the Tonga Trench and the Kuril Trench.

The Atlantic Ocean

The Atlantic has only about half the area of the Pacific, but in the north has a well-marked continental shelf adjacent to Europe and North America. The centre of the Atlantic is marked by a long ridge running roughly equidistantly between Europe and Africa, and the Americas. The ridge, sometimes called the Atlantic Rise, is interrupted at the equator by the Romanche Trench and supports the islands of the Azores, Ascension and Tristan da Cunha. The existence of this ridge adds weight to the theory of **continental drift** which holds that Europe, Africa and the Americas were once joined together and drifted apart to form the depression now occupied by the Atlantic.

The Indian Ocean

The Indian Ocean differs from the Atlantic and Pacific in that it lies mainly to the south of the equator and is enclosed by land to the north. It, too, has a ridge which stretches from the tip of India to Antarctica. The greatest deeps are in the north-east adjacent to the islands of Indonesia.

The waters of the oceans

Ocean waters vary considerably in temperature and in composition, especially with regard to the mineral substances contained both in solution and in suspension. They also vary enormously in the amount of marine life which they contain.

Salinity

There are a number of mineral salts present in sea-water but sodium chloride, or common salt, is the most important. Variations in salinity depend on the supply of fresh water, rapidity of evaporation and changes caused by mixing. In the open ocean the greatest salinity is found around the Tropics of Cancer and Capricorn owing to rapidity of evaporation. Towards the equator, however, heavy rainfall and low evaporation owing to cloudiness lead to lower salinity. Towards the poles melting ice supplying fresh water causes dilution and lowers salinity. Local variations in salinity are especially evident near coasts owing to the supply of fresh water from rivers of varying size.

Although some enclosed seas such as the Baltic have a relatively low salinity, most have a high mineral content. The salinity of the Mediterranean increases from west to east. Inland seas and lakes often have a very high salinity, for while their rivers may bring down only a small quantity of salt, it becomes concentrated there as a result of water being removed by evaporation.

Ocean currents
Considerable movements of water take place within the oceans both vertically and horizontally. The density of sea-water depends on both salinity and temperature. Vertical movements are caused either by differences in density at various depths or by the meeting of two currents, which causes sinking and must be counter-balanced elsewhere by upwelling.

The general movement of a mass of surface water horizontally in a fairly defined direction is known as an **ocean current**. The pattern of the oceanic circulation is produced by the interplay of a number of factors. Density differences are important but many of the surface currents are '**drifts**' caused by the friction between winds and the surface water. They therefore move more or less in the direction of the wind and vary seasonally in position and strength. Moreover, rotation of the earth, besides affecting the direction of the winds themselves, tends to deflect the currents slightly obliquely. In addition the shape of the land-masses helps to determine the direction. The terms 'warm' and 'cool' are applied to the various currents, describing their temperature relative to the surrounding seas and atmosphere. In general, a poleward-moving current is warm while an equatorward-moving one is cool. These currents may have a considerable effect on the climates of bordering land-masses.

Coral formations
In tropical seas a major feature of many coasts is the presence of coral. Corals exist in many forms. Some, known as **polyps**, live as individual animals but most are joined to each other in great

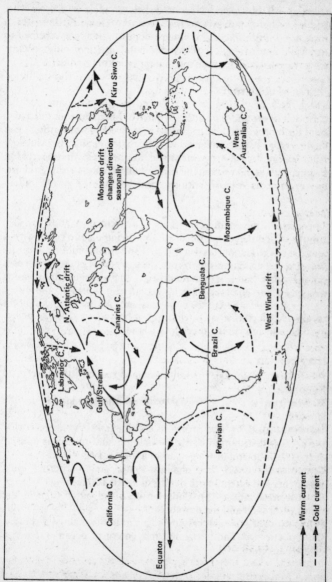

Figure 36. Major ocean currents

93

colonies. All, however, have a hard skeleton of calcium carbonate, and, as each polyp dies, a mass of skeletal material accumulates to form coral limestone. Coral limestone formations are confined to tropical or near-tropical seas since coral can live only in clear oxygenated salt water with a temperature of more than 20°C.

Types of coral reefs

Coral reefs may form around the edges of continents such as Australia or around the shores of islands. Independently, they can also form low coral islands rising apparently from the ocean depths, such as the Gilbert and Ellice Islands and the Marshall Islands. Coral is most widespread in the western and central Pacific, but is also found in the Indian Ocean. In the Atlantic, however, it is almost entirely confined to the West Indian archipelago.

There are three main formations. The **fringing reef** consists simply of an uneven platform of coral fringing the coast with a long, shallow lagoon between it and the mainland, and with its seaward edge sloping down into deep water. The **barrier reef**, on the other hand, is separated from the mainland by a much deeper, wider channel. The largest in the world is the Great Barrier Reef off the coast of Queensland. An **atoll** consists of a coral reef, which is circular, elliptical or horseshoe-shaped, enclosing a lagoon and having no central island.

Key terms

Atoll A ring-shaped or horseshoe-shaped reef enclosing a central lagoon.
Barrier reef A coral reef separated from the mainland by a wide lagoon.
Continental drift The theory of continental drift suggests that the present distribution of the continental land-masses is a result of the break-up and drifting apart of a former larger land-mass.
Continental shelf The shallow water area bordering most continents and extending to depths of about 100 fathoms.
Continental slope The slope between the outer edge of the continental shelf and the deep oceans.
Fringing reef A coral reef lying adjacent to the mainland shore.
Ocean current The horizontal movement of ocean water in a defined direction.

Chapter 12
Weather

The term 'weather' is used to denote the conditions of the atmosphere at a particular place and time. It normally includes reference to conditions of temperature, precipitation, pressure, humidity, cloud cover and wind. **Weather** refers to conditions at a specific point in time and should not be confused with the term **climate** which has been described as the average or normal weather for a particular place or region. Although weather is changeable and often unpredictable, it is nevertheless possible to arrive at a generalisation or average of these variations, which is referred to as climate.

Weather-recording

It is an obvious but important point that students attempting questions on weather-recording should have had some experience of carrying out the appropriate observations and measurements involved. Poor examination answers are inevitably produced by students who have never seen or handled the various instruments used for weather-recording.

Many schools have their own simple meteorological station and keep records of local weather conditions. An essential feature of any such weather-station is a **Stevenson screen** which is used to house many of the instruments. This is a white-painted wooden box or cabinet with louvred sides (so that the air can circulate freely round the instruments), standing on legs about 1 metre above the ground. It should be positioned on open ground away from the shade of trees or buildings. Instruments, such as the rain-gauge, which are not kept in the screen, are normally positioned adjacent to it. Elements of weather which are normally measured and recorded are listed below.

Temperature
Temperature is measured by means of **maximum and minimum thermometers** kept in the Stevenson screen. (Be able to describe and sketch these instruments.) Measurements are taken each day, normally at 9 a.m., and the highest and lowest temperature for the preceding twenty-four hours are noted. The instruments are then reset. The averaging of the maximum

temperatures for each day of the month gives an average monthly maximum temperature. Similarly, an average monthly minimum temperature can be calculated. The mean monthly temperature is obtained by calculating the mean value of the average temperature figures for each day of the month. Mean monthly temperatures published in atlases, etc., are based on records kept over a long period of time – usually at least thirty years. Temperature data are recorded on maps by means of **isotherms** (lines joining places of equal temperature). Normally the figures are adjusted to a sea-level equivalent (6·5°C for every 1,000 metres) before the map is drawn. This enables comparisons to be made between different areas without the influence of altitude overriding all other factors.

Precipitation

Precipitation is measured by means of a **rain-gauge**. This is a cylindrical container, 12 centimetres in diameter, with a funnel top which leads the rainwater into a removable flask. The total rainfall is measured and recorded every twenty-four hours. In order to obtain accurate measurements the rain-gauge must be sited on open ground well away from buildings and trees. The total rainfall for a month is obtained simply by adding the totals for each day of the month when rain was recorded. Mean monthly rainfall is obtained by averaging the same month's totals over several years. Mean annual rainfall is found by averaging several years' totals. Rainfall maps are constructed by using **isohyets** (lines joining places receiving equal amounts of rain).

Pressure

Atmospheric pressure is measured by means of a **mercury or aneroid barometer**. Pressure is expressed in millibars, the average pressure at sea-level being 1,012 millibars. Pressure is shown on maps by means of **isobars** (lines joining places of equal pressure). Normally readings are reduced to sea-level equivalents so that comparisons may be made between places of different altitude.

Relative humidity

Two thermometers, known as **wet and dry bulb thermometers**, are used to measure relative humidity. The wet bulb thermometer has its bulb of mercury kept moist by a strip of gauze, one end of which is placed in a dish of water. Air contains water vapour. The amount it can contain varies according to the temperature; the higher the temperature, the more water vapour it can contain. When the air contains the maximum water vapour possible for a given temperature it is said to be saturated. This temperature is the **dew point** for that air. When air is saturated, no moisture will be evaporated from the damp gauze surrounding the wet bulb thermometer, and the two thermometers will show the same reading. If the air is not saturated, evaporation will take place from the wet bulb thermometer. Evaporation requires heat, and the wet bulb thermometer will show a lower temperature than the dry bulb. The drier the air, the faster the evaporation and the greater the difference between the wet and dry bulb thermometer readings. Relative humidity, which is expressed as a percentage (completely wet air = 100 per cent relative humidity), can be found by referring the two temperature readings to a published table.

Wind speed and direction

Wind speed is measured by a **cup-anemometer** which consists of three metal cups attached to a revolving shaft at a height of 10 metres above the ground. The result may be given in knots or kilometres per hour or expressed as a Beaufort Number (see page 103). Wind direction is shown by a **wind vane**. Note that winds are always named according to the direction *from* which they blow.

Sunshine

This is recorded by traces burnt on to a sensitised card by the sun through a glass sphere. The instrument is known as a Campbell-Stokes **sunshine recorder**. A brown trace is scorched on to the card which is calibrated in hours, and the duration of sunshine, whether continuous or interrupted by cloud, can be read off. The recorder must be set up in an open situation away from buildings and trees.

Cloud cover

As well as the instrumental measurements described above, a visual assessment is normally made of cloud conditions. Cloud cover is expressed in oktas or eighths. This is the fraction of the sky covered by cloud. It is also normal to subdivide the cloud into high, medium and low cloud. High clouds of cirrus, cirro-cumulus and cirro-stratus extend to heights of 12,000 metres, medium cloud of alto-cumulus and alto-stratus up to 6,000 metres, and low cloud of strato-cumulus, stratus, nimbo-stratus, cumulus and cumulo-nimbus ranges from ground level up to 2,000 metres. A leaflet (No. 716) entitled 'Cloud Types for

Observers' and published by the Meteorological Office provides a useful guide to the recognition of the various cloud types.

Factors influencing temperature

The earth receives heat from the sun by radiation. Much heat is lost by reflection from dust particles and water droplets in the atmosphere. The amount of heat absorbed also depends on the nature of the earth's surface. Ice and snow, for example, reflect about 90 per cent of the sun's heat energy, compared with rocks, which reflect only about 14 per cent. Water surfaces are also heated to a greater depth than land surfaces. The earth's surface is heated directly by the sun; the lower layers of the atmosphere are heated by conduction from the earth's surface, and the upper layers of the atmosphere by convection. The temperature of any area is affected by the following factors.

Latitude

The higher the angle of the sun is above the horizon, the greater the heating effect of the sun's rays. There are two reasons for this: first, the sun's rays have a shorter path through the atmosphere, and secondly, they have a smaller surface area to heat up (see figure 37). Hence, the nearer a place is to the equator, the more heat it will receive. This situation is, however, complicated by the varying length of daylight experienced at different latitudes. High-latitude areas, for example, have low sun angles, but during the summer the long hours of daylight compensate for this fact.

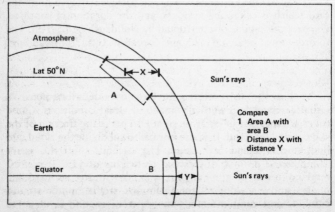

Figure 37. Temperature and latitude

Distance from the sea

Land heats up more quickly and cools down more quickly than the sea. Thus, continental interiors, remote from the moderating effects of the sea, have very hot summers and very cold winters. Areas with this type of large temperature range are described as having an **extreme** or continental climate. By contrast, coastal areas, especially those with onshore winds, generally have cool summers and mild winters, and are described as having an **equable** or maritime climate. The point can be well illustrated by comparing the temperature figures of two Canadian towns, Victoria and Winnipeg, both on approximately the same latitude, but the one situated on the coast and the other far inland.

Victoria   Lat. 49°N   January   4°C   July 16°C
Winnipeg Lat. 50°N   January   −18°C   July 19°C

Even a coastal area will have an extreme climate if its prevailing wind is offshore. This will have the effect of moving air masses from inland areas over the coastal area concerned. Thus, eastern Canada and northern-eastern USA have an extreme type of climate.

Altitude

As mentioned earlier, the atmosphere is heated from the earth's surface rather than directly by the sun, so that the higher the altitude of an area, the lower is its temperature. The average rate of decrease of temperature with height, known as the **lapse-rate** of the atmosphere, is by 6·5°C for every 1,000 metres rise. Thus, high-mountain and plateau areas have substantially lower temperatures than adjacent plains.

Ocean currents

The main ocean currents shown in figure 36 should be learnt. Their effect on the temperature of adjacent land areas is most pronounced where they are crossed by prevailing onshore winds. For example, in north-west Europe onshore westerly winds crossing the North Atlantic Drift bring unusually warm conditions for the latitude to the west coast of Norway in winter. Another interesting illustration is to compare temperatures of the south-west and south-east coasts of Africa. The former is washed by the cold Benguela Current and the latter by the warm Mozambique or Agulhas Current. Thus, Port Nolloth (Lat. 29°S) on the south-west coast has a mean January temperature of 16°C, while Durban (Lat. 30°S) on the south-east coast has a mean January temperature of 25°C. Mean July temperatures are 13°C and 18°C respectively.

Cloudiness

The effect of clouds is to reduce radiation from the sun to the earth, and also from the earth to the atmosphere. Thus, regions experiencing a frequent and persistent cloud cover tend to have small annual ranges of temperature. On the other hand, regions characterised by clear, cloudless skies tend to have large annual ranges of temperature. Cloud cover also affects diurnal temperature ranges. In tropical deserts, for example, the absence of cloud allows extremely high temperatures to develop during the daytime, but at night the heat loss from the land is rapid and temperatures may even fall below freezing point.

Aspect

The aspect or orientation of slopes in relation to the sun has a significant influence upon their temperatures. In the northern hemisphere south-facing slopes receive greater amounts of direct sunshine than their north-facing counterparts, and consequently enjoy higher temperatures. The effects of aspect on temperature are well demonstrated in deeply-cut, east–west trending valleys such as the Engadine in Switzerland. In such valleys the cool, shady, north-facing slopes are frequently wooded and devoid of settlement, while the warm, sunny, south-facing slopes are usually cultivated or used for pasture, and provide sites for village settlements.

Local winds

Local winds, as opposed to the major planetary winds, affect only relatively small areas. However, their effects on temperature and humidity may be very marked. For example, the **Föhn** wind which affects the northern Alps, and is caused by the warming of descending air from the south, causes temperatures to rise by several degrees within the space of a few hours, reduces humidity, and often causes avalanches owing to the rapid melting of snow. A number of well-known local winds are listed below.

Name	Characteristics	Area affected
Bora	Cold	Adriatic coast (Yugoslavia)
Chinook	Warm, dry	Alberta (Canada)
Föhn	Warm, dry	The northern Alps
Harmattan	Hot, dry	Sahara to the Guinea coast
Khamsin	Hot, dry	Egypt
Mistral	Cold	Lower Rhône valley (France)
Sirocco	Hot, dry	Southern Italy

Rainfall

As mentioned earlier, the amount of water vapour that a given air mass can contain varies according to its temperature. Warm air can contain more water vapour than cold air. If air is cooled below the temperature at which it is saturated with water vapour (**dew point**) then some of the water vapour will be condensed as water droplets around minute particles of dust, smoke, salt, etc., in the atmosphere. These water droplets form cloud, or if at ground level, fog. If cooling and condensation continue, the water droplets will grow until they are large enough to fall as rain. Thus, all processes of rainfall formation involve cooling of the air and the reduction of its temperature to below dew point. Three methods of cooling are commonly recognised, and on this basis three types of rainfall are distinguished.

Relief or orographic rainfall

When winds meet a barrier of high land they are forced to rise. As the air rises it expands (owing to the reduced pressure at higher altitudes) and cools, and may be reduced to dew point. Thus, clouds and rain may be produced. On the leeward side of the high land, the air is descending, and is warmed by compression. It is therefore able to absorb water vapour. Consequently, the leeward sides of mountain ranges are much drier than the windward sides. These dry areas are known as **rain-shadow** areas. For example, Patagonia on the leeward side of the Andes receives under 250 millimetres of rain per year while parts of southern

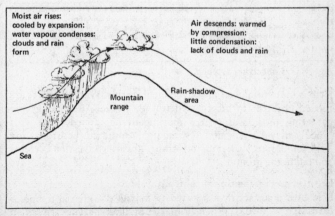

Figure 38. Relief or orographic rainfall

101

Chile on the windward side receive over 5,000 millimetres per year. For similar reasons Fort William on the windward side of the Scottish Highlands has 1,998 millimetres of rain per year compared with Aberdeen on the leeward side with 836 millimetres per year.

Convectional rainfall

Land heated by the sun in turn heats the air in contact with it by conduction. The heated air rises and its place is taken by cooler air. As the heated air rises it expands and cools. Dew point may then be reached and water vapour condensed to form clouds and possibly rain. The clouds commonly associated with convection currents are of the cumulus type. Convectional rain may be experienced in almost any region after intensive heating of the land, but is particularly associated with continental interiors in the summer and equatorial regions at all times of the year.

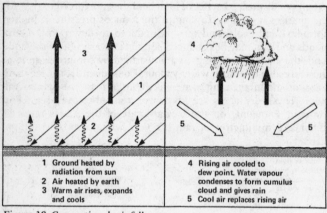

1 Ground heated by radiation from sun	4 Rising air cooled to dew point. Water vapour condenses to form cumulus cloud and gives rain
2 Air heated by earth	
3 Warm air rises, expands and cools	5 Cool air replaces rising air

Figure 39. Convectional rainfall

Hail is often produced in convectional storms. Water droplets are carried upwards to heights where they freeze. After falling and gathering more water they may again be carried aloft. This process may be repeated several times until the hailstones are large enough to overcome the strong upward convection currents and fall to the ground.

Frontal or depressional rainfall

A front is a surface dividing two air masses of different temperatures (see page 103). When air masses of different types are drawn together, as in a depression, the warmer air, being lighter, will tend to rise over the colder air. Alternatively, as along a cold front, the cold air may drive a wedge under the warm air so lifting it off the ground. The lifting of the warm air will cause it to expand and cool, and clouds and rain may follow. Frontal rainfall is experienced particularly in mid-latitude regions where air from the polar high-pressure zones meets air from the sub-tropical high-pressure zones.

The weather map

In Britain weather maps are prepared by the Meteorological Office from reports of observations sent in from various weather stations. Isobars, drawn at intervals of 4 millibars, indicate areas of high and low pressure. The weather conditions at selected stations are shown by a series of symbols. These are shown on page 103 and should be learnt. Note too how the various symbols are arranged around each weather station shown on the map (the station model).

Questions on weather maps are frequently set on both GCE O-level and CSE examination papers. In many cases an actual weather map is presented and the student required to interpret the weather situation shown. Alternatively, questions may require a description and sketch map of one of the more common pressure patterns. These include depressions, troughs of low pressure, anticyclones, and ridges of high pressure.

Depression

A depression is a centre of low pressure indicated on the weather map by a series of roughly concentric closed isobars. In the northern hemisphere winds blow around the low in an anti-clockwise direction, but cutting slightly across the isobars towards the centre of low pressure. Since air of different types is drawn into the low, it follows that fronts will be formed between the different air masses. Most depressions travel across Britain in a general west to east direction. In the front of the southern part of a well-marked depression warm, humid, maritime air from the south-west advances to meet colder, drier air from the south-east. The warm air rises over the cooler air along a surface known as a **warm front**. In the rear of the depression, cool north-easterly or northerly air undercuts the warm south-westerly air. This surface is the **cold front** of the depression, and again warm air is lifted off the ground. The fronts represent zones where lifting and cooling of air is taking place. Therefore these are zones of cloud and rain.

WEATHER SYMBOLS

≡	mist	•	rain	⍜	rain shower
≡	fog	✳	snow	⬦	hail shower
،	drizzle	⟨	thunderstorm	✲	snow shower

CLOUD SYMBOLS

⊗	sky obscured	◑	4/8 cloud cover
○	cloudless sky	◑	5/8 cloud cover
◐	1/8 cloud cover	◕	6/8 cloud cover
◕	2/8 cloud cover	◕	7/8 cloud cover
◕	3/8 cloud cover	●	complete cover

THE BEAUFORT SCALE   WIND SYMBOLS

Beaufort no.	Description	Speed (knots)	Symbol
0	calm	0	◎
		1 – 2	
1	light air	3 – 7	
2	light breeze	8 – 12	
3	gentle breeze	13 – 17	
4	moderate breeze	18 – 22	
5	fresh breeze	23 – 27	
6	strong breeze	28 – 32	
7	moderate gale	33 – 37	
8	fresh gale	38 – 42	
9	strong gale	43 – 47	
10	whole gale	48 – 52	
11	storm	53 – 57	
12	hurricane	58 – 62	

FRONTS

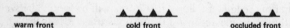

warm front   cold front   occluded front

(Symbols are placed on the side of the line towards which the front is moving)

STATION MODEL

Meaning: south-west breeze of 18–22 knots;
temperature 10°C; dew-point temperature
8°C; present weather rain; pressure
1020 mb; sky 6/8 cloud cover; past weather
drizzle.

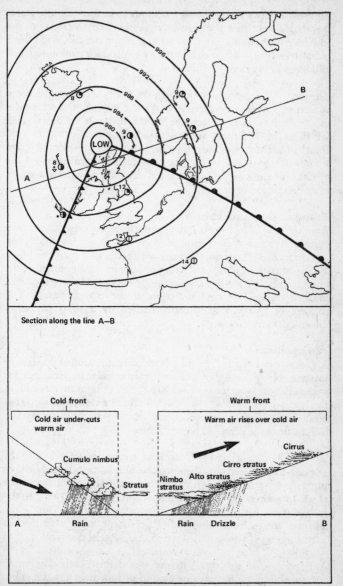

Section along the line A—B

Cold front

Cold air under-cuts warm air

Cumulo nimbus

Stratus

Warm front

Warm air rises over cold air

Cirrus

Cirro stratus

Alto stratus

Nimbo stratus

A   Rain   Rain   Drizzle   B

Figure 40. A depression centred over north-west Scotland

Figure 40 shows the sequence of clouds to be found as the warm front advances, from the high-level, feathery cirrus clouds to the low-level rain clouds of nimbo-stratus. Along the cold front there is more violent uplift of the warm air to produce cumulus clouds and often heavy rain.

The part of the depression between the fronts is referred to as the **warm sector** (the air there is warm and humid); the remainder of the depression is termed the **cold sector**. Most depressions which pass over Britain are **occluded**. This means that the cold front, which advances most rapidly, has caught up with the warm front and the whole of the warm sector has been lifted off the ground. In the case of an occluded depression there is a single extended period of rainfall.

Trough of low pressure

Another type of low-pressure system frequently experienced over Britain is known as a trough. Very often this is simply an extension of a depression. On the weather map it is marked by a V-shaped pattern of isobars with pressure increasing towards the point of the V. Wind speeds tend to be strong and there is often a marked change in wind direction as the trough passes. The passage of a trough is usually marked by a short period of unsettled, rainy weather.

Anticyclone

An anticyclone is a region of high pressure surrounded by areas of lower pressure. Isobars form a roughly circular pattern, but are generally widely spaced. Winds are therefore light or variable. In the northern hemisphere winds circulate in a clockwise direction around the high pressure centre, cutting slightly across the isobars from the centre of high pressure.

The weather associated with anticyclones varies according to the season. In summer they bring calm, hot, sunny days, but cool nights owing to heat being radiated from the earth into clear skies. In winter, days are calm and bright, but cold. Owing to the lack of cloud cover, hard frosts are common at night. During the autumn anticyclones often bring persistent foggy weather. In such conditions **smog** may develop in industrial districts.

Anticyclones do not travel in any well-defined path. They may remain stationary for days or expand or retreat in any direction. Changes are usually slow. Summer anticyclones are frequently terminated by convectional thunderstorms generated by the intensive heating of the land surface.

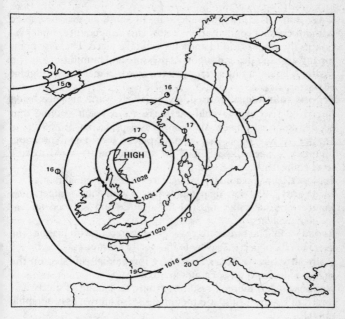

Figure 41. An anticyclone centred over the North Sea

Ridge of high pressure
This is a roughly triangular area of high pressure with pressure highest at the centre and lowest towards the point and sides. A ridge of high pressure is very often an extension of an anticyclone, or may simply be an area of relatively high pressure between two depressions. The weather is usually similar to that of an anticyclone, fine and with light winds, but lasts for only a relatively short time.

Key terms
Anticyclone An area of relatively high atmospheric pressure. In the northern hemisphere winds blow in a clockwise direction out from its centre.
Convectional rainfall Rainfall caused by the intensive heating of the earth's surface which causes air to rise and cool so that its moisture condenses to form rain.

107

Depression An area of relatively low atmospheric pressure. In the northern hemisphere winds blow in an anticlockwise direction towards its centre.

Dew point The lowest temperature to which air may be cooled without causing condensation. Below this temperature condensation in the form of cloud and rain will take place. The dew point for any air mass depends upon its pressure and humidity.

Front A boundary or surface between two air masses of different characteristics.

Frontal rain Rainfall caused by a relatively warm air mass being forced to rise over a cold air mass. As a result cooling and condensation take place. Also known as depressional rainfall.

Isobar A line on a map joining places with equal pressure. Figures are normally adjusted for altitude and reduced to sea-level equivalents.

Isohyet A line on a map joining places receiving equal rainfall.

Isotherm A line on a map joining places with equal temperature. Figures are normally adjusted for altitude and reduced to sea-level equivalents.

Lapse-rate The rate of decrease of temperature with height. The average rate of decrease is by $6\cdot5°C$ per 1,000 metres.

Rain-shadow An area of relatively low rainfall situated on the sheltered leeward side of a mountain range.

Relief rainfall Rainfall resulting from the cooling of air as it is forced to rise over hills and mountains. Also known as orographic rainfall.

Smog A type of fog experienced in industrial areas which contains a high proportion of smoke and dust particles. Sometimes known as smoke-fog.

Chapter 13
World Climate Regions

In order to explain the climatic conditions experienced in different parts of the world it is necessary to understand the main features of the world's pressure and wind systems. These should be carefully studied and learnt by reference to atlas maps showing features of world climate. In very general terms, two belts of high pressure are found at latitudes 30°N and 30°S, and two centres of high pressure located at the North and South Poles; low-pressure belts lie between these high-pressure zones, namely at latitudes 60°N and 60°S and the equator. Since winds blow from regions of high pressure to low pressure, the planetary wind-system can be explained in relation to the various pressure belts. In fact the various winds do not blow directly from the high-pressure belts to the low-pressure belts, but are deflected as a result of the earth's rotation. Thus, the north-east and south-east trade winds affect the Tropics, the north-west and south-west variable winds blow across middle-latitude regions, and the north-east and south-east polar winds blow from the polar regions towards the low-pressure belts at 60°N and 60°S. This simple pattern of pressure and winds is, in fact, complicated by two factors: the uneven distribution of land and sea over the earth's surface, and the seasonal movements of the overhead sun. These two influences will be examined in turn.

In summer, continental interiors are heated so strongly that centres of low pressure are developed. These tend to draw air into them. This effect is most marked in the case of the largest continent, Asia, where the **monsoon** winds completely displace the trade winds over the south-eastern part of the continent. In winter continental interiors are so cooled that high-pressure centres develop. Again the greatest effect is seen in south-east Asia where dry monsoon winds blow out from the land to the sea.

The various pressure belts described above also migrate in response to the seasonal movement of the overhead sun, but lag some time behind it. Thus, the equatorial low-pressure belt moves to about 5°N in July and 5°S in January. The effect of this seasonal movement of the pressure belts is most marked in areas experiencing a Mediterranean type of climate (see page 113). In winter these areas come under the influence of the onshore variable winds, but in summer experience high-pressure calms or offshore trade winds. Any attempt to divide the world into a series of distinctive climate regions poses many problems. Inevitably climatic conditions change gradually and imperceptibly from one region to another, and any boundary line is therefore very arbitrary. Furthermore, within any one region there will be local variations of climate in response to localised factors of relief, altitude, etc. A very basic division of the world into twelve climate regions is presented here. In each case the climate characteristics are described and some reference made also to the natural vegetation found in each region. **Figure 42 shows the location and extent of each of the regions described, and the abbreviations (TC, CTE, etc.) shown on it refer to the paragraph headings below (pages 110–17 inclusive).**

Equatorial climate (E)

Areas experiencing an equatorial climate straddle the equator, extending approximately to latitudes 5°N and 5°S. They include the Amazon basin, coastal Ecuador, the Congo basin, Malaysia and Indonesia. The noonday sun is high throughout the year and is actually overhead at the equator on the equinoxes, so that temperatures are constantly high. The annual range of temperature rarely exceeds 3°C, and no seasons can be distinguished. Annual rainfall is heavy (over 2,000 millimetres) and is of the convectional type, occurring daily throughout the year. In highland areas such as the East African Plateau and the Andes of Ecuador temperatures and rainfall totals are both lower, and conditions are more favourable for settlement.
Example: Singapore (height 31 metres).
January: 26·4°C; May: 27·8°C. *Temperature range*: 1·4°C.
Precipitation: 2,410 millimetres (all seasonal).
Owing to the heavy rainfall and constant high temperatures natural vegetation is of the **rain-forest** type. Giant trees of 50 metres or more form a canopy above more moderately-sized trees. Ground plants compete for light among a profusion of roots, fallen branches and lianas (creepers). Trees include mahogany, rosewood, ebony, greenheart and other hardwoods.

Tropical continental (Sudan) climate (TC)

Areas of tropical continental or Sudan climate are found on either side of the equatorial climate belt. They include the Brazilian Plateau, Orinoco Basin, Sudan, Rhodesia, Zambia, Tanzania, parts of northern Australia, etc. The climate is one of contrasts. The period of the overhead sun brings convectional rainfall and high temperatures (often over 32°C). The opposite season is characterised by dry conditions with high-pressure calms or offshore

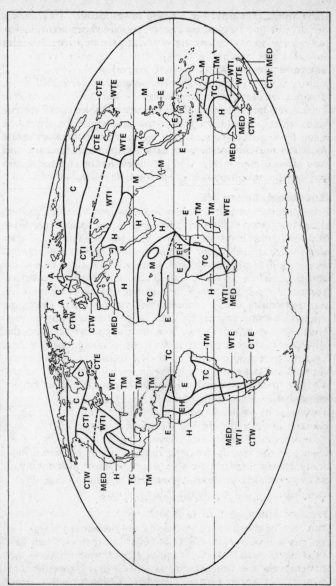

Figure 42. World climate regions

trade winds. Temperatures are also lower (about 21°C) during the dry season. Tropical continental climate varies according to position. Towards the equator rainfall totals are higher and the annual range of temperature less.

Example: Kayes, Mali (height 30 metres).
January: 25·0°C; May: 34·2°C. *Temperature range*: 9·2°C.
Precipitation: 760 millimetres (summer maximum).

Typical vegetation consists of tall **savanna grass** (up to 2 metres high) with scattered trees, such as baobabs, which are adapted to living through the dry season. Along river banks trees are more abundant and form **gallery forests**. In Africa and Australia this type of vegetation merges into rain forest on the equatorial side and into desert vegetation in the opposite direction.

Hot desert climate (H)

Areas of hot desert climate are found in or near the tropics, often along the western sides of continents. Areas include northern Chile and Peru (the Atacama Desert), southern California, north-western Mexico, the Sahara Desert, Saudi Arabia, Iran, north-western India (the Thar Desert), south-western Africa (the Kala-hari and Namib Deserts), and central and western Australia. The climate is characterised by very hot summers (the noonday sun is overhead at $23\frac{1}{2}°$ latitudes) and warm winters. Owing to the lack of cloud cover the diurnal range of temperature is very great. Annual rainfall is very low, for these areas lie under the influence of high-pressure systems or offshore trade winds. Even where onshore winds occur they usually cross cold ocean currents and therefore yield little rain, owing to the fact that the air is warmed as it moves inland and any water vapour rapidly evaporates.

Example: In Salah, Algeria (height 280 metres).
January: 13·3°C; July: 36·7°C. *Temperature range*: 23·4°C.
Precipitation: 15 millimetres (very sporadic).

Owing to the very high temperatures and lack of rainfall only drought-resistant plants are able to survive. Most have thorny or waxy leaves, deep roots or some means of water storage. Plants include thorn-bush, sage-bush, palms and cacti.

Tropical maritime climate (TM)

This type of climate is experienced along the eastern margins of continents between latitude $5°$ and $23\frac{1}{2}°$ (apart from South East Asia). Areas include the West Indies, Central America, the coasts of Venezuela, the Guianas, and south-east Brazil, together with eastern Malagasy, and coastal Queensland. Climate in these areas is dominated by onshore trade winds which bring hot, wet conditions at all times of the year. The annual range of temperature rarely exceeds 6°C.

Example: Havana, Cuba (height 24 metres).
January: 22·2°C; July: 27·8°C. *Temperature range*: 5·6°C.
Precipitation: 1,219 millimetres (all seasonal; slight summer maximum).
Natural vegetation consists of rain forests of hardwood trees similar to those of equatorial climate regions.

Tropical monsoon climate (M)

Tropical monsoon climate is experienced in South East Asia (India, Bangladesh, Burma, Sri Lanka, Thailand, Laos, Cambodia, Vietnam, and south China) and northern Australia. The climate is one of marked seasonal contrasts produced by changes in temperature and pressure over central Asia and Australia. In winter intense cold prevails over interior Asia owing to the rapid cooling of the land. This encourages the formation of a large high-pressure centre. Thus, in winter air moves from land to sea (the north-east monsoon of India), bringing dry conditions. In summer intensive heating of the Asian land-mass produces a deep centre of low pressure which draws winds onshore from the sea (the south-west monsoon of India), to give a season of very heavy rainfall. Thus, there is a marked seasonal reversal of winds and a sharp division of the year into a rainy season and dry season. Rainfall totals are influenced by position and relief. Thus, Cherrapunji in the hills of Assam which directly face onshore winds from the Bay of Bengal receives 10,795 millimetres of rain per year, and is reputedly the wettest place on earth. By contrast, Delhi on the Ganges Plain has only 625 millimetres of rainfall per year.

Example: Bombay, India (height 11 metres).
January: 23·9°C; July: 29·7°C. *Temperature range*: 5·8°C.
Precipitation: 1,808 millimetres (marked summer maximum).
Natural vegetation varies according to the amount of rainfall received. In the wettest areas there are evergreen forests similar to those of equatorial climate regions. Where rainfall is less, deciduous woodland prevails, the trees losing their leaves in winter to reduce transpiration. Species include teak, sandalwood, bamboo, banyan, etc. In the driest areas vegetation is restricted to thorny scrub and grassland.

Mediterranean climate (Med)

This type of climate, which is sometimes referred to as warm temperate west margin climate, is experienced along the western sides of continents between latitudes 30° and 40° North and South. Areas include the lands bordering the Mediterranean

Sea, central California, the Cape Town region of South Africa, central Chile, and the regions around Adelaide and Perth in Australia. Climate in these areas is determined by the seasonal movements of the pressure and wind belts as they follow the overhead sun. In winter Mediterranean climate regions come under the influence of onshore westerlies which bring depressional rain to give warm, wet winters. In summer they lie under high-pressure calms or in the path of offshore trade winds, so summers are hot and dry. Mediterranean climate may be regarded, therefore, as a transition between hot desert climate on the equatorial side and cool temperate west margin (British) climate on the polar side.

Example: Rome, Italy (height 34 metres).
January: 7·2°C; July: 24·5°C. *Temperature range*: 17·3°C.
Precipitation: 830 millimetres (winter maximum).
Natural vegetation varies according to the amount of rainfall. Mountain slopes which are relatively cool and wet support mixed or deciduous woodland (sweet chestnut, beech, plane, cypress, cedar, pine, etc.). Cork-oak and olive are found in lowland areas. Areas of low rainfall or poor soils support a cover of xerophytic or drought-resistant shrubs which include lavender, broom, myrtle, laurel, wild olive, etc. Various names are given to this poor scrub vegetation, including **maquis** (southern France), **garrigue** (France), **chaparral** (California) and **mallee scrub** (Australia).

Warm temperate east margin climate (WTE)

This type of climate, which is sometimes referred to as China or Gulf climate, is experienced in the same latitudes as Mediterranean climate, but on the eastern sides of the continental land-masses. Areas include south-eastern USA, southern Brazil, Uruguay, Natal, coastal New South Wales, south-eastern Victoria, central and northern China. Offshore variables cause dry conditions in winter (although depressional rain occurs in some areas), and onshore trade winds bring summer rain. Summers are generally hot (over 21°C), but winters may be quite cool with air masses moving from cold continental interiors.

Example: Charleston, USA (height 3 metres).
December: 10·9°C; July: 27·5°C. *Temperature range*: 16·6°C.
Precipitation: 1,209 millimetres (summer maximum).
Typical vegetation consists of woodland or forest dominated by species such as cedar, mulberry, eucalyptus and tree-ferns (in Australia).

Cool temperate west margin climate (CTW)

This is also known as British or temperate maritime climate, and is

found along the western sides of continents between latitudes 40°
and 60° North and South. Areas include the British Isles, north-
west Europe, coastal British Columbia and Alaska, southern Chile,
Tasmania, and South Island, New Zealand. Climate is characterised by warm summers and mild winters. The annual range of
temperature is moderate or small. Rainfall occurs at all seasons and
is the result of onshore westerlies and the passage of depressions
over the areas concerned. Variations in rainfall totals are caused by
relief. In many cases climate is influenced by warm ocean currents.
For example, winter temperatures in coastal north-west Europe are
raised by the North Atlantic Drift, and in Alaska and British
Columbia by the North Pacific Drift.
Example: Valentia, Eire (height 10 metres).
February: 6·9°C; July: 15·3°C. *Temperature range*: 8·4°C.
Precipitation: 1,420 millimetres (all seasonal).
Natural vegetation in most areas of cool temperate west margin
climate consists of deciduous or mixed woodland. Lowland areas,
which have now been largely cleared, were originally covered
with woodlands of oak, elm, ash, etc. Coniferous species are
found in upland areas and areas of poor soils.

Cool temperate east margin climate (CTE)

Cool temperate east margin or Laurentian climate, as it is
sometimes termed, is found along the eastern sides of continents
between latitudes 40° and 50° North and South. The St Lawrence
Valley, north-eastern USA, Patagonia, Manchuria, and northern
Japan all experience this type of climate which differs from the
west margin or British type of climate in respect of its far colder
winters. There are two reasons for this. Firstly, the prevailing
westerlies bring cold air from the continental interiors in winter,
and secondly, many of the areas mentioned above are influenced
by cold ocean currents such as the Labrador and Oya Siwo
currents. Thus, temperatures in the St Lawrence Valley, for
example, fall well below freezing point in winter. Rainfall, which
is all seasonal, is of the depressional type, although Manchuria
and northern Japan experience a monsoonal effect which produces rather drier winters than elsewhere.
Example: Toronto, Canada (height 116 metres).
February: −5·3°C; July: 20·6°C. *Temperature range*: 25·9°C.
Precipitation: 815 millimetres (all seasonal).
Natural vegetation in areas of cool temperate east margin climate
is generally a mixture of coniferous and deciduous woodland. In
Patagonia the low rainfall caused by the rain shadow effect of the
Andes results in a poor steppe or thorn-bush vegetation.

Temperate interior climate (Cool: CTI; Warm: WTI)

This type of climate is experienced in continental interiors in middle latitudes. Areas include the High Plains and Prairies of the USA and Canada, interior Argentina, Eastern Europe, central USSR, parts of South Africa, and the Murray-Darling plains of Australia. In these areas the land heats rapidly in summer, but loses its heat quickly in winter (see page 99). Thus, the climate is characterised by a very large temperature range. This is especially true of continental interiors in the northern hemisphere which lie far from the moderating effects of the sea: Winnipeg in central Canada, for example, has an annual temperature range of 37°C. In the southern hemisphere the continents are narrower and the annual range of temperature far less. Rainfall is light or moderate, occurring chiefly in summer in the form of convectional rainstorms.

Example: Edmonton, Canada (height 677 metres)
January: −14·7°C; July: 16·4°C. *Temperature range*: 31·1°C.
Precipitation: 439 millimetres (summer maximum).

The low rainfall occurring mainly in summer encourages the growth of grassland. This is known as **prairie** in North America, **steppe** in Eurasia, **pampas** in South America and **veld** in South Africa. Tree growth is generally restricted to water-courses.

Cold temperate climate (C)

Cold temperate climate is experienced in the northern parts of Eurasia and North America, stretching in a broad belt from the west to east coasts. In the centre and east of the continents it extends as far south as 50°N, but only to 60°N along the west coasts. This type of climate is not experienced in the southern hemisphere because the continents do not extend sufficiently far south. Climate is dominated by the high latitude. Summers are short and cool, and winters very cold, especially in the interior areas. Precipitation varies in amount and distribution according to position, being heavy and all seasonal along the west margin and light in the interior. In eastern Canada it is also all seasonal and heavy, but in eastern Asia is of a monsoonal type and confined to the summer months.

Example: Dawson City, Canada (height 324 metres).
January: −29·4°C; July: 15·3°C. *Temperature range*: 44·7°C.
Precipitation: 320 millimetres (slight summer maximum).

Natural vegetation is of a coniferous forest type, sometimes referred to as **taiga.** Softwood trees such as pine, spruce, larch and hemlock form some of the greatest uninterrupted stretches of forest in the world. Trees are adapted to the long cold-weather

116

drought, with needles taking the place of normal leaves and so reducing the surface area through which transpiration takes place.

Arctic or polar climate (A)

This type of climate is confined to the northernmost parts of Eurasia and North America. The southern limit is generally taken as the 10°C July isotherm; that is to say, temperatures very rarely rise above 10°C even in summer. Summers are short and cool, and winters long and extremely cold. Precipitation is light, occurring chiefly in summer when depressions may reach these northern latitudes.

Example: Jacobshavn, Greenland (height 37 metres).
February: −18·9°C; July: 7·5°C. *Temperature range*: 26·4°C.
Precipitation: 234 millimetres (summer maximum).

Natural vegetation is very restricted owing to the adverse conditions for plant growth. During the brief summer growing-season the ground may thaw out to a depth of about 50 centimetres below which it is permanently frozen (the **permafrost** layer). Typical vegetation consists of mosses, lichens, and dwarf species of willow, birch and alder. The term **tundra** is applied to this type of vegetation.

Key terms

Gallery forest Forest occurring along river banks in otherwise open country.

Maquis A scrub vegetation occurring in the Mediterranean climate region of France. Similar vegetation is known elsewhere as garrigue, chaparral, and mallee scrub.

Monsoon A wind characterised by regular seasonal changes in direction caused by the differences in temperature and pressure of adjacent land and sea areas. The term is applied in particular to the wind system affecting much of South East Asia.

Permafrost The permanently frozen subsoil of high-latitude areas.

Prairie Grassland of the North American continental interior regions. Similar vegetation is known elsewhere as steppe, pampas, and veld.

Savanna Grassland of tropical or subtropical regions containing only scattered trees.

Taiga The coniferous forest found in high-latitude areas of the northern hemisphere.

Tundra A level or gently undulating plain found in Arctic areas and characterised by a permanently frozen subsoil supporting a vegetation of mosses, lichens and dwarf shrubs.

Chapter 14
World Population and Food Supply

World distribution of population

In 1973 the total world population was estimated at 3,865 millions. Reference to atlas maps will show that the distribution of this population is extremely uneven, with vast areas of sparse population on the one hand contrasting with highly crowded areas supporting high population densities on the other hand. Approximately 80 per cent of the total population occupies less than 20 per cent of the earth's land surface.

The main points of contrast are between the northern and southern hemispheres, and between the Old World and the New World. More than 90 per cent of the world's population inhabits the northern hemisphere, and less than 10 per cent is found in the southern hemisphere. Similarly the Old World (Eurasia) supports over 85 per cent of the world's population, while the New World contains less than 15 per cent of the total. These discrepancies are caused in part by the inequalities of land area in the northern and southern hemispheres and in the Old World and the New World, but even if one takes this fact into account and considers population densities rather than population totals, the same contrasts are evident.

Three **primary concentrations** of population on a world scale are generally recognised; namely, South East Asia which contains approximately 50 per cent of the world's population, Europe with about 20 per cent, and north-east North America which contains a further 5 per cent of the total. Various **secondary concentrations** may also be noted. These are regions of high population density, but limited in extent. They include California, eastern Brazil, areas around the Plate Estuary, South Africa and south-eastern Australia. These secondary concentrations together account for about 5 per cent of the world total. Finally, there are a number of smaller **tertiary concentrations** of population such as the Nile Valley, the areas around Lake Victoria, and the high basins of Mexico.

In contrast to these densely-populated lands are vast areas of the earth's surface which remain only sparsely populated, or in some instances virtually uninhabited. These include high-latitude areas, especially the lands north of the Arctic Circle, desert and semi-desert areas, high mountain and plateau areas, and areas of dense equatorial forest.

The term **ecumene** is sometimes used to signify the inhabited areas of the earth's surface, and **nonecumene** to refer to the uninhabited or sparsely-inhabited areas. It has been estimated that approximately 60 per cent of the earth's land surface may be classified as ecumene and 40 per cent as nonecumene. However, it will be appreciated that these figures provide no more than a rough indication of the relative proportions, for it is by no means easy to draw a boundary between the ecumene and nonecumene. Areas of high population density merge gradually into low-density areas. Furthermore, the ecumene contains areas of sparse population, while isolated settlements and communities exist within the nonecumene. In any case the boundary between the two is not static. In earlier times an expansion of the settled lands was a characteristic feature of world history, but more recently the settlement frontier has shown signs of retreat in many marginal areas. At the present time there are relatively few areas of active population advance into empty lands.

Influences on population distribution

From the study of a map showing the world distribution of population it is tempting to conclude that physical factors such as extreme cold, aridity, and high altitude are the essential influences on population distribution. Although physical or environmental factors are important, they generally provide no more than a partial explanation of any population distribution, and various historical, political, social and economic factors must also be taken into account.

Among the physical influences on population distribution, **altitude** is clearly an important factor. It imposes an upper limit on human habitation, so that few permanent settlements are found above 5,000 metres. Approximately 56 per cent of the world's population lives between sea-level and 200 metres, which includes only 28 per cent of the total land area. Just over 80 per cent of the total population lives below 500 metres. Altitude is closely related to the factor of **relief** which also influences population and settlement patterns. Steep gradients and rugged terrain deter settlement by restricting movement and the possibilities for cultivation, while river valleys and lowland plains generally support high population densities. **Climate** is another factor which is thought to influence population distribution, but its role is difficult to evaluate. It is often argued that the climate of cool and warm temperate margins provides the ideal stimulus for human settlement and economic activity. On the other hand, the largest primary concentration of population, accounting for half the world total, is found in a region of monsoon climate, which can hardly be regarded as the most attractive climate with its seasonal contrasts of heavy rain and drought and constant high temperatures. Extremes of cold and aridity clearly deter settlement, but the ideal climate is difficult, if not impossible, to define. **Soils** indirectly affect patterns of population by their influence on agriculture. Deltaic and alluvial soils generally attract agricultural populations, while naturally infertile soils such as laterites and podsols normally support only a sparse population. In a similar manner, variations in **natural vegetation** provide contrasting environments for different types of agricultural activity, which in turn are capable of supporting different levels of population density. The distribution of population in many countries is also affected by the location of **mineral resources**. For example, the population map of Western Europe reflects to a very large extent the distribution of coalfields and their associated mining and industrial towns. In other instances, such as northern Canada or the interior of Australia, mineral deposits have attracted settlements far beyond the limits of the ecumene.

In addition to these purely physical factors, attention must also be paid to various social and economic influences on population patterns. Much depends, for example, on the **age of settlement**. The low population density of Australia (2 persons per km²) can be largely explained in terms of its relatively recent settlement, whereas the high density of India (170 persons per km²) stems partly from its long history of settlement and civilisation. Differences in age of settlement largely account for the contrasts in population density between the Old World and New World noted earlier. The **type and scale of economic activity** practised in any region is another important factor. The Industrial Revolution brought about striking changes in the population pattern of many countries. Pre-industrial, agricultural populations tend to be fairly evenly distributed, whereas in highly industrialised countries population has been attracted to coalfields and ports, and dense concentrations of population have replaced former patterns of dispersal and even distribution. In recent times **political influences** on population patterns have become increasingly important. In the communist world, population may be directed to areas of social, economic or strategic need, while in the western world, governments offer various incentives to attract population away from overcrowded urban centres to new towns and special development areas. As throughout history, political events continue to be responsible for enforced mass movements of refugees in many parts of the world.

World population growth

At the present time virtually every nation of the world is characterised by an increase in the size of its population. During the present century world population has increased at a rate unprecedented in the history of mankind, so that world population growth is now one of the most urgent problems of our time.

The 1973 world population total of 3,865 millions may be compared with figures of 2,486 and 2,982 millions for 1950 and 1960 respectively. In recent years the increase in world population has been at a rate of approximately 70 million persons per year. This figure is equivalent to a population larger than that of the UK or about one-third that of the USA. It represents a monthly addition to the world's population equivalent to a city ten times the size of Leeds, or a daily addition equal to the population of Southampton or Derby.

This present rate of population growth is a relatively recent phenomenon. It seems clear that a spectacular acceleration of population growth dates back to the mid-seventeenth century. The annual rate of population doubled between 1650 and 1850, doubled again between 1850 and 1920, and yet again between 1920 and 1970. A United Nations Population Bulletin in 1951 pointed out that 'it took 200,000 years for the world's human population to reach 2,500 million; it will now take a mere 30 years to add another 2,000 million'.

World hunger and malnutrition

As early as 1798 Thomas **Malthus**, in a now famous book *Essay on the Principle of Population*, drew attention to the fact that population tends to increase at a faster rate than the increase in food supply and the means of subsistence. He suggested that population has a tendency to increase in a geometric ratio (1, 2, 4, 8, 16, 32, . . .), while food production tends to increase in an arithmetic ratio (1, 2, 3, 4, 5, 6, . . .), the former being subject to

121

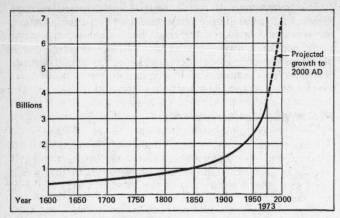

Figure 43. World population growth

population checks such as wars, famine and disease, and the latter subject to improvements in food-supply owing to technical developments in agriculture.

During the nineteenth century vast areas in Russia and the New World were opened up for cereal production and ranching, and many improvements took place in the transportation and distribution of foodstuffs. As a result, the pessimistic views and fears of Malthus seemed unfounded, and his ideas were largely neglected. However, with the dramatic acceleration of world population growth during the present century, many people have reconsidered his ideas and applied them to the contemporary population problem.

In recent years a great deal has been written about the relationship between world population growth and the world's present and potential food-supply. The viewpoints and interpretations are varied. On the one hand, there are those who point with dismay at the present levels of hunger and **malnutrition** in the world, note the increasing regularity of regional **famine** disasters, and predict a deterioration of the present situation as the world's population continues to increase at its present alarming rate. On the other hand, there are those who hold a more optimistic view, believing that large-scale birth-control programmes can check the present excessive rate of population increase, and suggesting that many opportunities exist for increasing the world's food-supply.

122

At the present time about half the world's population is under-nourished, and the greater part of mankind lives at an extremely low standard. At North American living standards the world could support a population of only 500 million people; at European standards about 1,500 million; yet the world's population has already reached 4,000 million. It was recently stated that at least one in ten of the present population do not even get the 1,900 or so calories a day needed just to maintain the structure of their bodies. Other food deficiencies hurt in every possible way, from the 100,000 Far Eastern children who go blind every year through lack of vitamin A to the Latin American communities where half the children suffer from significant anaemia.'

Although less dramatic than the regional outbreaks of famine, malnutrition is an equally serious problem, impairing health and vitality, and producing **deficiency diseases** among a very large proportion of the world's population. Deficiency diseases are caused by two main dietary gaps, protein–calorie malnutrition and vitamin–mineral deficiencies. They include kwashiorkor, anaemia, trachoma, pellagra, beri-beri, goitre and many other diseases. Furthermore, malnutrition lowers resistance to other diseases and infections including tuberculosis, leprosy and malaria. It has been estimated that over half the world's population has suffered at some time or other from one or more of these diseases which reduce energy and the ability to cope with both physical and intellectual effort, qualities which are most urgently needed in the developing world where these diseases are most widespread.

The possibilities of increasing world food production

It has been estimated that approximately 60 per cent of the earth's land surface is totally unsuitable for cultivation on account of extreme cold, aridity, high altitude or lack of soil. At present about 15 per cent of the land area is under cultivation, leaving a further 25 per cent which offers some possibilities for an expansion of the farming frontier.

This **expansion of agriculture** into new lands could take many forms. Clearance of equatorial forest could yield additional farm-land, though past experience has shown that land gained in this way requires very careful management if its fertility is to be maintained. Commercial ranching has been successfully intro-duced into a number of tropical grassland areas, and there is every

123

indication that this type of development could be greatly extended. Large-scale irrigation schemes can bring agricultural prosperity to low-rainfall areas, while in middle latitudes the drainage of marsh and peat areas and the reclamation of heathland could add to the existing farmland. In high-latitude areas the limits of cultivation could be further extended by the development of new strains of cereals which require a shorter growing period. Another form of agricultural extension also exists: namely, the creation of new areas of farmland by the reclamation of tidal swamps and shallow offshore areas. This technique has been perfected in the Netherlands where 40 per cent of the country has been reclaimed from beneath the sea. Thus, many possibilities for agricultural expansion exist, but new land can only be brought into production at very high cost, and capital must be available, probably in the form of financial aid from the economically advanced nations of the world, if significant gains of new farmland are to be achieved.

Many possibilities also exist for an **intensification of agriculture** and an increased yield of food from those parts of the earth's surface which are already under cultivation. Better management and improved farming techniques could check and reduce soil erosion in many parts of the world. Greater use of fertilisers could significantly increase crop yields. Techniques of plant breeding such as hybridisation could be used to develop new strains of plants which are more resistant to disease, mature and ripen more quickly, and produce higher yields per unit area. In a similar manner, selective stock-breeding can be used to rear high-quality livestock which are more resistant to disease, and yield increased amounts of milk and meat. At the present time a great deal of food is lost to a variety of pests and diseases which affect both crops and livestock. A greater use of insecticides, fungicides and herbicides (weed-killers), and the spraying and vaccination of livestock would do much to reduce these losses, although some concern is now being shown about certain adverse effects resulting from the excessive use of neurotoxin insecticides such as DDT. Increased mechanisation of farm-work is yet another means of increasing agricultural efficiency and output.

Many of these technical changes require supporting programmes of education, as well as changes in social organisation. Excessively small farm holdings must be amalgamated into economic units to make investment in machinery worthwhile. The advantages of new farming methods must be clearly demonstrated on model farms if outdated, traditional practices are to be abandoned in favour of new methods. Improvements in production must also be matched by modernisation of the methods of food storage, processing and distribution, in order to avoid the present high losses, deterioration and wastage which occur in many parts of the world as a result of inadequate facilities.

Finally, it is worth noting the future possibilities of extracting protein from a variety of unusual sources such as seaweed, plankton, organisms such as bacteria and yeast which can be cultivated under controlled conditions, and the residual materials left after oil has been extracted from soya beans, cotton seed, sunflower seed, and other sources of vegetable oils. It has also been estimated that the present yield of food from the sea could be increased by at least 50 per cent by the development of techniques of 'fish-farming' which are currently in an experimental stage.

Key terms

Deficiency disease A disease resulting from malnutrition and chronic undernourishment, being caused by protein–calorie deficiencies or vitamin–mineral deficiencies.

Ecumene The settled or permanently inhabited parts of the earth's surface.

Malnutrition A deficiency of food or an unbalanced diet which is inadequate for normal body requirements. Chronic undernourishment.

Natural population increase The excess of births over deaths. Ignores additions and losses resulting from migration.

Nonecumene The uninhabited or temporarily inhabited parts of the earth's surface.

Population density The ratio between population numbers and land area. Usually expressed as the number of persons per km^2 or per mile^2.

Chapter 15
Agriculture

Agriculture or farming may be defined as the growing and collecting of crops (tillage and harvest), and the care and use of animals (husbandry) to provide food products and/or industrial raw materials. The type of agriculture practised in any area is the result of many influences. Physical factors (climate, soil and relief) determine the broad characteristics of the agricultural pattern, but in most environments there is a wide range of crops and livestock able to thrive under the conditions present. Thus, the actual selection of crops and livestock in any area is also the result of social and economic influences, either past or present. Other things being equal, farmers generally produce the crops or livestock which are in greatest demand and yield highest profit.

Physical influences on agriculture

The physical influences on agricultural production fall into three basic groups: climate, soils and relief. The elements of **climate** which influence agriculture include rainfall (the total amount and its seasonal distribution and intensity), evaporation rates, amounts of sunshine, wind speed, and temperature conditions. The latter are particularly important. Plant growth normally ceases when temperatures fall below about 5°C, so that the length of growing season, or period with temperatures above 5°C, is of critical importance. In high-latitude areas the length of growing season is too short for many crops. The occurrence of frost during the growing season can cause serious damage and crop losses. **Soil** is the medium in which seeds germinate and crops grow and ripen, and, as such, exercises a basic influence on agriculture. The features of soil which affect its potential for crop growth include its depth, composition and texture, as well as the depth of the water-table. For many crops, especially cereals, deep, well-drained clay-loams provide the ideal growing medium, but other crops are less demanding and will give satisfactory yields on soils of moderate quality. Various elements of **relief** also affect agriculture: notably, altitude, gradient and aspect. Owing to the reduction of temperature with altitude, the length of growing season diminishes with increased height. In Britain, for example, most upland areas are given over to rough grazing and pastoral farming. Steep slopes restrict the use of heavy machinery and may necessitate the construction of terraces in order to bring them under cultivation.

Aspect or orientation of slopes influences agriculture through its control on temperature (see page 100).

Social and economic influences on agriculture

The type of farming practised in any area is influenced not only by the physical factors described above, but also by a number of social and economic considerations such as farm size, system of land tenure, market demand, availability of labour, transport facilities, tariffs and other government influences, etc. For example, in areas where farms are small and the land fragmented into small plots, the farmer has no alternative to intensive production methods which give a high return per hectare. Rice cultivation in monsoon Asia typifies this situation. On the other hand where land is cheap and readily available and farms are large, extensive methods of arable farming or ranching are more suitable, as, for example, on the Canadian prairies or the Argentinian pampas. Different crops and livestock vary in their labour requirements. Where labour is costly or in short supply it may be necessary to adopt highly mechanised farming methods, provided that capital is available for investment in machinery. In recent years government influences on farming have also become very significant. By imposing tariffs and quotas on agricultural imports and by offering guaranteed prices and subsidies for particular crops, national governments are able to influence the pattern of production within their own countries. Finally, mention should be made of the role of the individual farmer. His assessment of what crops should be grown may not necessarily be the correct or best evaluation for his particular farm. In making his decisions he may be influenced by local traditions, or incomplete knowledge about the range of alternatives available. For these reasons the actual pattern of farming in any area may not match up with the potential pattern offered by the physical environment.

Farming systems

Two basic approaches may be adopted in the study of the variations in world agriculture. Firstly, the different systems or **methods of organisation** of farming activity can be examined, and, secondly, the emphasis can be placed on the **type of farming**, focusing on the distribution of specific crops and livestock.

Farming systems can be divided into two broad categories: namely, **subsistence agriculture** in which crops are grown for local use, and **commercial agriculture** in which crops are grown for sale or exchange. Subsistence agriculture is generally characterised by rudimentary, traditional methods of farming with a low level of mechanisation, while commercial agriculture frequently employs modern, scientific methods to maintain soil fertility and improve yields, and is often capital-intensive and highly mechanised. Various systems are described briefly below:

Shifting agriculture Communities practising shifting agriculture are found in equatorial, monsoon and savanna areas where plant growth is quick and easy. In forest areas, trees are felled, undergrowth burnt, and cassava, yams, etc., planted. Little work is done during the growing season. Soil fertility soon declines because of leaching, and the community then moves to another part of the forest. This way of life is typified by the Boro of Amazonia.

Nomadic herding Nomadic herders rely almost entirely upon the products of their animals. For example, the Lapps of Arctic Scandinavia rely upon their reindeer for meat and milk, skins for clothing, bones for tools and ornaments, as well as a means of transport. Most domesticated animals are herbivores, and therefore most pastoral communities are found in the grassland areas of the world such as the tropical savannas (the Masai of East Africa), temperate steppes (the Kazak and Kirghiz of Turkestan), and Arctic pastures (the Lapps of northern Scandinavia). Desert pastoralists, such as the Bedouins of Arabia, use oases to provide water for their animals. Because they must continually search for new pastures or water-supply, the various groups mentioned above lead a **nomadic** existence. The term **transhumance** refers to seasonal movements of men and livestock.

Tropical peasant agriculture In systems of tropical peasant agriculture the types of crops grown and the methods of farming vary from one region to another, but in general, farms are small and worked by family labour. Subsistence crops are grown rather than cash crops. Profits are small and little or no capital is available for farm improvements or mechanisation. Crops grown include rice (in wetter areas) and millet (in drier areas). **Rice cultivation** is in many ways the most characteristic form of tropical peasant agriculture, and rice forms the basic food for over 50 per cent of the world's population.

Plantation agriculture The term 'plantation agriculture' describes the large-scale production of cash crops (usually from tropical areas) for sale in temperate market areas. Plantations involve local labour and European or US management and capital, and they represent the organisation of tropical crop production on scientific lines together with the economies of large-scale production. However, notable failures have occurred when European farming techniques have been employed in the tropics without modification. Crops include rubber, coffee, sugar-cane, cocoa, cotton and tobacco.

Intensive mixed farming This type of farming is typified by agricultural activity in many parts of Western Europe and the north-eastern USA. Farms are often small, owner-occupied, but characterised by high returns per hectare. The types of crops grown and the livestock raised vary from area to area. Wheat is frequently the dominant food grain, with barley, oats and rye in less favourable areas. Root crops such as swedes, turnips and mangolds are grown for sale or winter feeding of livestock. Industrial crops such as sugar-beet, flax, oil-seeds, etc., are important in some areas. In wetter areas intensive **dairy farming** may be an important specialised local activity. Sheep are generally raised on tracts of poorer land. In many countries such as Denmark, Finland, the Netherlands, etc., highly organised forms of **co-operative farming** have developed.

Market gardening Market gardening is one of the most highly specialised forms of agricultural activity, concerned with the production of high-value 'speciality crops' for urban markets. It is generally carried out close to the urban areas served, as, for example, in the Lea Valley north of London, or in areas with outstanding climatic or other advantages such as the Scilly Isles and Channel Islands. Production units are small, but intensive production methods give an extremely high return per hectare. Crops such as lettuce and cucumbers are often grown under glass.

Extensive arable farming In many of the New World countries together with the steppe-lands of the USSR a distinctive form of arable farming has evolved. This is characterised by very large farms, relatively little manual labour, but a high level of capital investment in machinery. Yields per hectare are low compared with those of the intensive farming systems of Western Europe. Large areas are devoted to the cultivation of a single crop (**mono-culture**), usually wheat. In the USSR and much of Eastern Europe agriculture is state-run (**collective agriculture**).

Extensive ranching The rearing of livestock on very large ranches is another distinctive type of agriculture in many New World countries. Sheep-farming in Australia, the raising of beef cattle and sheep in Argentina and Uruguay, cattle-ranching on the interior grasslands of the USA and Canada, all typify this type of farming. Much of the meat produced in these countries is exported to West European countries, a situation made possible by the development of refrigerated ships in the 1870s. Ranches are extremely large, but support only a small number of animals per hectare compared with areas of pastoral farming in Western Europe.

Crops and livestock

As mentioned on page 127, another approach to the study of agriculture is to examine the distribution of specific types of crops and livestock. In this approach reference is normally made to the physical requirements and tolerances of each crop, its main production areas, volume of production, and importance in world trade. Listed below are details of some of the most important agricultural commodities. The various crops and livestock have been divided into two main groups: namely, those of tropical regions and those of middle-latitude or temperate regions. It should be appreciated that such a division is somewhat arbitrary. Wheat, for example, is listed as a temperate crop, but is also grown in relatively small amounts in a few tropical areas. Cotton is grown in both tropical and temperate areas. Similarly, India has more cattle than any other country in the world, but produces negligible amounts of meat, therefore it has been considered more appropriate to list cattle under temperate commodities.

Tropical crops

Most farming activity in the tropics is of a subsistence type. However, certain cash crops are often grown on plantations.

Rubber Natural rubber is produced from the sap (latex) of the rubber tree, which is native to the Amazon basin. It needs hot (over 25°C), wet (over 2,000 millimetres of rainfall) conditions. In 1876 plants were taken to Malaysia and Sri Lanka, and plantations established in those areas. At the present time natural rubber faces severe competition from the synthetic product. World output is 3·0 million tonnes per year, chiefly from Malaysia, Indonesia, Thailand and Sri Lanka.

Coffee Coffee is produced from the seed (bean) of the coffee shrub which requires hot (over 21°C), wet (over 1,500 millimetres of rainfall) growing conditions, but a sunny harvest period. Soils should be well drained. The shrub is frequently grown under trees which provide shade. World output is 5·2 million tonnes per year, chiefly from Brazil, Colombia, Ivory Coast, Angola, Mexico and Uganda.

Cocoa Cocoa is obtained from the seeds of the cocoa tree which requires hot (over 21°C), wet (over 1,250 millimetres rainfall) conditions, but shade from strong direct sun. The tree is very susceptible to a number of diseases and regional crop failures are not uncommon. World output is 1·5 million tonnes per year, mainly from Ghana, Nigeria, Brazil, Ivory Coast and Cameroon.

Tea Tea is produced from the leaves of an evergreen shrub native to South-East Asia. It likes hot, damp conditions and grows best on well-drained hillsides with deep soil. Total world production is 1·3 million tonnes per year, coming chiefly from India, Sri Lanka, China, Japan and Indonesia.

Sugar-cane Sugar-cane is a tropical grass which grows to heights of 2 metres. It requires hot (over 20°C), wet (over 1,250 millimetres of rainfall) conditions, and flourishes in tropical lowland areas. World output is 40 million tonnes per year, mainly from India, Brazil, Cuba, Bangladesh, China and Mexico.

Cotton Cotton is produced from a small shrub with seed-pods which contain cotton lint. It is a sub-tropical plant, and extremely sensitive to frost. It requires 200 frost-free days, with summer temperatures of 22°C and an annual rainfall of 500–1,000 millimetres. It is often grown under irrigation. Annual world output is 12·2 million tonnes, chiefly from the USSR, USA, China, India, Brazil, Pakistan and Egypt.

Rice Rice is an annual 'grass' which grows in water. A distinction can be made between hill (terraced) rice and paddy (swamp) rice. It requires rich alluvial soils with impervious material underneath it to retain the water during the growing season. It also needs high temperatures (over 21°C) and heavy rainfall (over 1,000 millimetres). Under favourable conditions two or even three crops per year are possible. Much hand labour is required and the work of planting, transplanting and harvesting does not lend itself to mechanisation. Total world production is 306 million tonnes per year, but a mere 5 per cent enters world trade. Chief producers are China, India, Indonesia, Bangladesh, Japan and Thailand.

Middle-latitude (temperate) crops and livestock
Wheat Although it is grown in a variety of regions, both temperate and tropical, wheat requires deep, fertile, well-drained soils in order to give high yields. It also needs a frost-free growing season of 90–100 days, three to four months over 13°C, and an annual rainfall of 400–1,000 millimetres. Irrigation or dry farming (see page 134) may be practised in areas of lower rainfall. World output is 353 million tonnes per year. Major wheat-

growing countries include the USSR, USA, China, Canada, France, India, Turkey, Italy, Argentina and <u>Australia</u>.

Barley Barley is cultivated both for human consumption and as a fodder crop. Its climatic and soil requirements are less demanding than those of wheat. World output is 150 million tonnes per year, chiefly from the USA and Europe.

Oats Oats prefer cooler, damper conditions than wheat. Oats will also give good yields on less rich soils. World output is 57 million tonnes per year.

Rye Like oats, rye prefers cooler, damper conditions than wheat. In Europe it is frequently grown in areas of relatively poor, sandy soils. World output is 32 million tonnes per year.

Maize Maize or corn is a native cereal of North America which is grown both for human consumption and as a fodder crop. It requires hot summers (25°C), 140 frost-free days, and an annual rainfall of 600–1,000 millimetres. It yields particularly well under conditions of warm temperate interior climate, as for example in the mid-western states of the USA. World output is 306 million tonnes per year, almost half of which is grown in the USA. Relatively small amounts are also grown by China, Brazil, South Africa, Mexico, the USSR and Argentina.

Potatoes Potatoes can be grown under a variety of conditions, but give the best yield and are of the best quality in areas with a cool, moist climate. World production is 293 million tonnes per year, with the USSR the leading producer. Since potatoes do not keep well they are unimportant in world trade.

Sugar-beet Sugar-beet requires deep, loamy soils, a summer temperature of 12°C–20°C, and an annual rainfall of 500–900 millimetres. Besides yielding sugar, the pulp and tops provide cattle fodder, and alcohol is obtained as a by-product from the refining process. World output is 30 million tonnes per year, chiefly from the USSR, USA, Poland, West Germany, Italy and France.

Cattle Cattle may be raised for either beef or milk production or as dual-purpose animals. Dairying is carried out in areas with equable temperatures and abundant rain. Important breeds of **dairy cattle** include Ayrshires, Guernseys, and Friesians. Countries important for the production of dairy produce (milk, butter, cheese, etc.) include the USA, USSR, New Zealand, Denmark, the Netherlands, Poland, France and the UK. In many dairying areas pigs are fed on skimmed milk from the dairies and reared for bacon production. Ranching of **beef cattle** takes place in drier areas such as the Great Plains of the USA or the South American pampas. Breeds of beef cattle include the Aberdeen Angus, Hereford, and Shorthorn. World production of beef totals 36 million tonnes per year, chiefly from the USA, Argentina, Uruguay, Australia, New Zealand, the USSR and UK.

Sheep   Sheep-farming is generally associated with drier or poorer grassland areas than those used for cattle, although sheep may be transferred to good-quality pasture for fattening prior to marketing. Sheep may be reared for lamb or mutton production or for wool, or as dual-purpose animals. The chief sheep-rearing countries of the world include Australia, the USSR, New Zealand, China, Argentina, Uruguay, the USA and the UK.

Farming problems and techniques
Soil erosion

Soil erosion may be defined as the destruction and removal of topsoil by either wind or water action. Soil erosion is generally the result of bad land management. It has been said that 'it takes nature from 300 to 1,000 years to build up one inch of fertile soil. Man by his wanton misuse can destroy 8 inches in one or two generations.' The main causes include the destruction of forests which regulate the surface run-off of rainfall, inadequate attention to the maintenance of soil fertility, over-cropping, over-grazing, etc. Soil erosion occurs in all continents. Nor is it restricted to developing countries. In the USA over half the land is affected by soil erosion in varying degrees.

Techniques of prevention and cure include strip-cropping, terracing of steep hillsides, greater use of fertilisers, contour ploughing, diversification of farming, programmes of afforestation, planting windbreaks or shelter-belts of trees, and making farmers aware of the causes and consequences of soil erosion.

Techniques of land improvement

Land drainage   The agricultural potential of areas of low-lying, waterlogged land can be improved by various techniques. These include the digging of ditches, the installation of tile- and pipe-drains in fields, the straightening of river courses to improve flow, the raising of river banks and the construction of levées to prevent flooding, and the building of sluices and pumping stations to control river discharge. In these ways badly drained land can be improved or even reclaimed for production, as, for example, in the Fens of eastern England, and the Dutch Polders.

Soil improvements Just as man is capable of initiating soil erosion by bad land management, so, by careful farming practices, is he able to improve the quality of the soil in any area. Similarly, areas of poor, infertile soils can be reclaimed and brought under cultivation, though this requires considerable time and capital investment. This involves the regular application of natural or artificial fertilisers to provide the minerals and trace-elements necessary for successful plant growth. **Crop rotation** also helps maintain soil fertility. Cultivating a single crop (monoculture) tends to deplete soil fertility, but a carefully organised rotation of different crops enables plant foods taken from the soil by one crop to be replaced by others. Certain nitrogenous legumes such as clover are particularly beneficial.

Dry farming Various techniques are used to retain soil moisture in low rainfall areas. These include cropping the land in alternate years only, and the application of a mulch cover (grass or straw) to reduce water losses by evaporation.

Irrigation

Irrigation techniques are generally practised in areas with less than 500 millimetres of rain per year, especially if the distribution is seasonal. The large modern schemes with huge river-dams and enormous reservoirs are frequently multi-purpose involving flood control, navigation improvement, hydro-electric power production, as well as irrigation. The different methods of irrigation used in various parts of the world include rudimentary bucket and water-wheel hoists such as the shaduf of Egypt, systems of tank storage as used in parts of India, artesian wells (as in Australia), basin irrigation involving the channelling of seasonal flood waters, and modern barrage and reservoir schemes such as that created by the Aswan Dam on the River Nile.

Key terms

Crop rotation An ordered sequence of crops on the same land designed to avoid soil exhaustion.

Monoculture The continuous cultivation of a single crop.

Nomadism The practice of moving from place to place, especially in search of pasture for livestock.

Shifting agriculture A primitive form of agriculture. Plots are cultivated until the soil is exhausted and then abandoned.

Transhumance The seasonal movement of livestock.

Chapter 16
Fishing and Forestry

Two of man's oldest economic activities are fishing and the exploitation of forest and woodland resources. Moreover, even with the vastly changed technology of the twentieth century, both continue to have a vital role to play.

Fishing

The chief fishing grounds of the world are mainly in the cool waters of high or middle latitudes and particularly where there is a mixing of warm and cool ocean currents. It is in such conditions that **plankton**, the microscopic plants and creatures upon which most fish feed, is abundant. The greatest numbers of fish are to be found in the shallow waters and offshore banks of the continental shelf, such as Dogger Bank in the North Sea.

As a form of economic activity fishing varies greatly in scale, from the large-scale industry carried on in ports such as Grimsby and Aberdeen to the small-scale activity for subsistence or local needs which is common in underdeveloped countries such as India. The growth of a commercial fishing industry depends on a number of factors. The availability of good natural harbours suitable for boat-building and the accommodation of a fishing fleet is vital. Fish is a highly perishable commodity and must be delivered to the consumer as quickly as possible. Where the large centres of population which are the chief markets for fresh fish are situated well inland, good road or rail links must, therefore, be provided. Otherwise, if commercial fishing takes place at a great distance from the market, the catch must be processed. **Processing industries** are now a feature of many fishing ports and take the form of canning, drying, smoking, salting and freezing of fish and the production of fish-oil and fish-meal.

Methods of fishing

The fishing industry is becoming increasingly scientific in its methods for detecting, catching and transporting the fish upon which it is based. Refrigerated vessels have allowed distant fishing grounds to be increasingly exploited. Recently, the use of echo-sounding has enabled the whereabouts of shoals of fish to be determined much more precisely.

The method of fishing depends on the type of fish being sought. In this connection the distinction between **pelagic fish**, such as herring, mackerel and pilchards, which live near the surface, and **demersal fish**, like cod, haddock and plaice, which live near the sea-bed, is important. There are a number of contrasting methods. **Drifting** is mainly used for pelagic fishing, for instance by the North Sea herring boats, and involves the suspension of a long net from floats. **Trawling** is used in deeper water, for example in the Icelandic cod industry, and involves a scoop-shaped net. **Long-line fishing** can be adapted for pelagic or demersal fish at almost any depth and is used in the New-foundland cod fisheries. **Seine-netting** involves the drawing of a long net across an estuary or the mouth of a fjord. It is especially useful for catching those fish, like salmon, which migrate inland from the sea, and is intensively used for salmon-fishing in British Columbia. **Inshore fishing** of the littoral zone for crabs, lobsters and a wide variety of shell-fish involves a number of different methods. **Fresh-water fishing** is sometimes commercially important. The salmon and sturgeon fisheries of the USSR provide a good example.

Today the possibilities of fish-farming are being increasingly explored. As fishing methods become more efficient the dangers of over-exploitation and the need to conserve and manage resources are becoming more apparent.

The chief fishing grounds of the world

Most coastal waters yield some fish but the areas where intensive, commercial fishing takes place are fairly localised. Most of these are located in the northern hemisphere, often in seas bordering densely populated areas. The continental shelf of the north-east Atlantic, the Grand Banks of Newfoundland and the seas around Japan and the coast of China come into this category. The Icelandic fishing grounds, Peruvian coastal waters and the Pacific coast of North America are also important. The chief fishing nations of the world are as follows:

Major fishing nations: catch landed (millions of tonnes)

Peru	10·6	India	1·8
Japan	9·9	Thailand	1·5
USSR	7·3	Spain	1·5
China	6·8	Denmark	1·4
Norway	3·0	Canada	1·2
USA	2·7	**World total**	**69·0**

Whaling

Whaling is an important industry in some countries since a single whale can be processed to produce large quantities of oil and animal foodstuffs. In 1972 the world catch was 30,683 whales and two countries, Japan (13,890) and USSR (10,286), were by far the most important. There is a real danger of over-exploitation of stocks and a great need for **conservation**.

Forestry

Although, in the past, wood had a much greater variety of uses, and materials like steel, plastics and fibre-glass have tended to supersede it, it is still an essential raw material for the paper, furniture and construction industries. In the densely populated parts of the world the natural forests have been rapidly depleted and a number of countries have embarked upon ambitious **afforestation** schemes in order to replenish their resources. In Britain, the Forestry Commission has been responsible for planting large forests in Scotland, Wales and northern England. However, Britain still needs to import most of her timber from Sweden and Finland.

Tropical hardwoods

These consist mainly of broad-leaved evergreen trees which yield hard, heavy timber. A few species, like balsa, however, produce a soft, light wood. West Africa, Brazil, India and parts of South East Asia export commercial quantities of species such as mahogany, rosewood, greenheart, ebony and teak. There are many problems in obtaining tropical hardwood, however, since much of it is still worked from natural forest. The unhealthy climate, especially the intense humidity, experienced in tropical rain forests inhibits prolonged manual labour. The dense undergrowth, with few roads or railways, makes penetration difficult. The large size and weight of the logs means that they are often too heavy to float down rivers and restricts overland transportation.

Unfortunately, the great length of time which it takes for most tropical hardwoods to mature makes it difficult to justify their use in afforestation schemes. Consequently, most tropical hardwoods continue to be obtained from scattered clearings near to main roads and railways. The chief uses for hardwoods are in the manufacture of expensive furniture and veneers.

Temperate hardwoods

These include those broad-leaved trees which grow in warm and cool temperate climate areas. A variety of different species are used, including oak, beech and ash in Europe, the quebracho in South America, the eucalyptus in Australia, the kari in New Zealand, the maple in Canada, the redwood in California and the cork-oak in Mediterranean areas. Like their tropical counterparts, the temperate hardwoods are chiefly used in the making of furniture and veneers, and in panelling.

Temperate softwoods

These are coniferous trees, such as pine, spruce and fir, which are found mainly in cold regions. Conifers make up most of the vast **boreal forests** or **taiga** which extend across northern Canada and Siberia. Since undergrowth is sparse and the trees often occur in uniform stands covering wide areas, conifers are much easier and cheaper to work than hardwoods. Most conifers mature fairly rapidly and are tolerant of a wide range of climatic and soil conditions so that they are ideal for **afforestation.** Many are now being planted in areas which were once covered by broad-leaved trees.

A regular pattern of working has been adopted in eastern Canada and Scandinavia. The trees are felled mainly in autumn and winter and the logs are stacked alongside the frozen rivers. In the spring the timber is floated on rivers swollen by melting snow to sawmills, ports or rail termini for processing or dispatch. Consequently, it is the timber near to the rivers which is the first to be utilised, and there are serious transport problems involved in the opening up of new timberlands away from the rivers.

Softwoods are used mainly in construction, furniture manufacture and the making of woodpulp, newsprint, paper and cardboard. The by-products of pulping include cellulose which is later converted into rayon and alcohol.

Timber production

The USSR, with its vast, untapped forest resources, and the USA dominate world production of both hardwoods and softwoods. World production of coniferous timber is 335 million tonnes, of which the USSR produces about 100 million tonnes and the USA 75 million tonnes, followed by Japan, Canada and Sweden. The USA and the USSR are also the main producers of broad-leaved timber. Out of a total world output of 93 million tonnes, they produce 17 million and 15 million tonnes respectively. The USA and Canada are the main manufacturers of woodpulp and paper.

Other forest products

As well as providing timber and food in the form of fruits, forests also supply many other products including tar, dyes, rubber, tannin, copal, lac, chicle, turpentine, resin, quinine, camphor, pitch, raffia and bamboo. It should be borne in mind, however, that in primitive societies timber has an even wider range of domestic uses, including that of fuel.

Key terms

Afforestation The replanting of trees in forest areas cleared for timber, or the planting of trees in former waste areas.

Conservation The protection or preservation of natural resources, including wildlife, from wasteful or destructive uses. A reduction in the rate of consumption or exhaustion of resources for the benefit of posterity.

Demersal fish Fish which live on or near the sea bed.

Pelagic fish Fish which live near the surface of the sea.

Plankton A collective name for the minute organic life, both plant and animal, found drifting at shallow depths in the ocean. The material upon which fish feed.

Chapter 17
Mining

The mining industry is a primary **extractive industry.** By its very nature mining is an exhaustive industry so that locational patterns change through time as deposits become depleted. The products of mining are used as raw materials for secondary industries. These products, which include both metallic and non-metallic minerals, are of great importance to modern manufacturing industry.

Occurrence of minerals

Minerals are found throughout the world, although many deposits are insufficiently rich or concentrated to warrant exploitation. Typically they are found in three main forms. **Bedded ores**, as the name implies, are found in beds or strata of varying thickness. Minerals commonly found in this form include iron and coal. **Veins or lodes** are the product of igneous or metamorphic action. Copper, tin, lead and zinc are generally found in this form. **Alluvial or placer deposits** are the result of the erosion of either bedded or vein deposits by rivers which then redeposit the eroded material as sediment.

Exploitation of mineral deposits

The exploitation of mineral deposits depends on a variety of factors other than the mere presence of minerals. The **quality** of the ore itself is important, as high-quality ores are more likely to be profitable than low grade or lean ores. This factor is largely controlled by the costs of transport and mining, as unit costs of recovered ore will be higher for the leaner ores. However, the quality of the ores which are mined constantly changes in response to changes in demand and technology. The effect may be seen in iron ore exploitation where techniques of beneficiation (concentration of ores at source) and pelletisation of the improved ores have allowed leaner ores to be used profitably, particularly in the USA.

Accessibility of the ore is also important, since unless the ore can be moved to market it has no commercial value. This factor also changes through time as new transport facilities are developed to allow exploitation, but as a rule the most accessible ores are exploited first. The question of accessibility may be illustrated by the Swedish iron ore industry. The rich iron ores of Kiruna and Gällivare in north Sweden had to await exploitation until the construction of a railway to Narvik in Norway, as the ports of north Sweden are closed by ice for almost half the year.

The **ease of mining** is also important. This is determined by factors such as relief and climate in the area of the deposits, and also their depth. The ease of exploitation affects different minerals in different ways according to their characteristics and value. Perhaps the most extreme example is in the mining of gold in South Africa where the depth of the lodes containing the gold is very great. The value of the product is such that the capital investment needed to work in such extreme conditions of depth and heat can be met. The working of other lower-value minerals in such difficult situations would not be profitable.

Deposits of ores are found in different quantities; some deposits are large, some small. The **size of the deposit** influences exploitation largely through the economies of scale which may be achieved. Small deposits are generally more expensive to exploit than large deposits, and consequently unit costs of production are higher. Larger deposits will tend to be exhausted before smaller deposits are worked.

Market demand for a mineral is also crucial to exploitation, and both influences the volume of production and determines which of several possible sources will be developed. Examples will illustrate this point. The growth in the use of iron and steel in the twentieth century has vastly increased the demand for iron ore. As demand has risen new reserves have been developed, reserves which were not previously exploited because of access difficulties, distance from markets, or the low quality of the ores. Similarly, the nature of demand may change as in the substitution of copper for lead in pipework, which created an increased demand for copper at the expense of lead and thus caused problems in lead-mining areas while stimulating the search for new copper reserves. The recent substitution of aluminium for copper in electrical applications has similarly increased the demand for bauxite at the expense of copper. These and other complexities of demand have a profound influence on the exploitation of minerals and the distribution of mining activity.

Closely linked to the demand factor is **technological change**; new uses are found for existing minerals, and uses found for new minerals. From the beginning of the Industrial Revolution this has been characteristic of the mining industry. The Gilchrist-Thomas process of steel-making gave new value to phosphoric iron ores such as the Cleveland ores in the UK, whilst the development of alloy steels for special uses, such as high-speed machining and corrosion resistance, have increased the demand for minerals such as zercon, tungsten, molybdenum and vanadium. Technological developments in the mining industry itself have led to a reassessment of many low-grade deposits, as methods of on-site concentration have been developed and improved.

Several other factors also influence the exploitation of mineral reserves, including **labour-supply, availability of water, capital reserves,** and the **level of economic development** of the country in which the reserves are found. The interaction of some or all of these factors determines the pattern of exploitation of mineral deposits in any area.

There are several **methods** by which minerals may be mined. These include deep mining and quarrying or opencast mining. In **deep mining** a shaft is sunk and the minerals are blasted out in underground galleries and raised to the surface. Since production costs increase with depth, it follows that very deep mines are only used for high-value minerals. Some gold mines in South Africa are over 3 kilometres deep. East Rand Proprietary Mines near Johannesburg, for example, extract gold at 3,200 metres. At these depths the technical and human problems are immense, and with rock temperatures at about 49°C a vast amount of refrigerated air must be pumped into the mine. Where minerals lie close to the surface they are exploited by **opencast mining** or **quarrying**. This is particularly important for lower-value minerals. The capital costs of quarrying are relatively low. For alluvial deposits, techniques not unlike the panning methods of the old-time gold miners are used. For example, in the recovery of tin ores in Malaysia huge dredges dig up the tin-bearing gravels from lake bottoms and pass the gravel over a series of troughs into which the heavier tin ore is deposited.

Different minerals vary very widely in their degree of concentration in the mined ore, and many require processing at the mine prior to transportation. Most iron ore requires no concentration at the point of extraction since the waste material or **gangue**, as it is termed, is removed during smelting by the addition of limestone in the blast furnace which converts it into slag. However, in the case of very low-grade iron ore, especially if it is to be transported over long distances, concentration may be carried out in the mining area. Copper and lead ores which may have less than 1 per cent metal content inevitably require concentration at the mine. A variety of different methods is used for concentration, including those which make use of the different densities of the mineral and gangue. When a current of water is passed through the ore the heavier minerals may be separated out from the gangue. Of particular importance is the **flotation process**. By this method the dense minerals are made to float in a bath of 'frothed' liquid, while the unwanted matter sinks. Finely crushed ore is agitated in water containing one or more chemical frothing agents. The gangue is usually wetted and sinks, while the minerals rise in the froth and are skimmed off and dried. This method is being increasingly used as growing demand for many minerals necessitates the use of poorer quality ores.

The major minerals (excluding fuel minerals)
Iron
Iron ore is the most important of the non-fuel minerals both in terms of overall demand and the volume entering world trade. The most important types of iron ore are magnetite and haematite, both high-grade ores containing up to 70 per cent iron. Limonite and siderite, both lower-grade ores, are also mined in many areas. The major world producers of iron ore are tabulated below.

Major iron ore producers, 1972 (millions of tonnes)

USSR	113·5	Canada	24·4
USA	45·8	Liberia	22·5
Australia	39·2	India	22·2
Brazil	28·6	Sweden	21·3
China	25·3	**World total**	**430·0**

At the present time the world's known workable reserves are about 275,000 million tonnes, enough for over 200 years at present rates of consumption. There are also proven reserves in excess of 550,000 million tonnes which are not economically viable with present methods of extraction. Known reserves are added to annually as new discoveries are made.

The USSR is by far the largest producer of iron ore, most of which is used internally and within Eastern Europe. It also has the world's largest iron mine at Lebedin near Kursk which has reserves of 20,000 million tonnes of rich iron ore. Many other large deposits also exist, as for example at Murmansk near the Finnish border. The rise of Australia as a major world producer is related not only to the increased demand of Australian manufacturing industry but also to the huge growth of the Japanese steel industry (see page 162). Old-established mining centres such as Iron Knob and Iron Monarch in South Australia supply internal demand, the ores being shipped from Whyalla to Newcastle. More recent mining developments such as that at Yampi Sound in Western Australia supply overseas markets, principally those in Japan. Enormous ore-carriers are used to transport the ore on a long-term contract basis. The UK produces only about 45 per cent of the 35 million tonnes of ore needed annually. Home production comes chiefly from Northamptonshire and Lincolnshire. The average iron content of these ores is only about 27 per cent which is low by international standards. The remaining 55 per cent of UK requirements is imported from Canada, Venezuela, Australia, Liberia, Mauritania, Brazil and Sweden.

Trade in iron ore in recent years has been dominated by the development of large, purpose-built ore-carriers which have considerably reduced the tonne-kilometre cost of transporting ore. This has made the long-distance movement of ores economic on a large scale. The Japanese are currently developing ore-carriers in excess of 200,000 tonnes. These carriers require deep-water berths and have necessitated the building of deep-water ports specifically designed to handle them. Narvik and Kirkenes in Norway, Port Cartier and Sept Iles in Canada, Puerto Ordaz in Venezuela, Port Monrovia in Liberia and Port Etienne in Mauritania are examples of such developments. Another recent development in ore shipment is found in relation to the black sand deposits of North Island, New Zealand. These sands at Tahatoe, south of Auckland, contain iron oxide and are used by the Japanese iron and steel industry. The sands are pumped into ore-carriers as slurry. The slurry is de-watered for the voyage and then converted back into slurry to be unloaded. The whole process is very highly mechanised so that over 1 million tonnes can be handled each year by as few as fifty men. Further developments in ore transportation are likely to ensure even larger carriers and more refined and mechanised methods of handling the ores.

Copper

Copper is found in a wide variety of geological formations throughout the world. Its uses are based on its properties of softness and ductility, electrical conductivity, and its ability to be used in alloys such as brass, bronze and duralium. Its major use is in the electrical industry, although the rapid rise in copper prices in world markets since 1970 has meant a decline in many electrical applications. The table below gives details of the major world producers of copper.

Major copper producers, 1972 (thousands of tonnes)

USA	1,510	Canada	709
USSR	1,050	Zaire	412
Chile	723	Peru	217
Zambia	717	**World total**	**6,425**

In the USA copper is mined in the Rocky Mountain states of Montana, Utah, and Arizona. Large-scale opencast methods are used. The USSR has major centres at Monchegorsk, Mednogorsk, Koundrad, Leninogorsk, and Amalyk. Many other reserves are known, particularly in Siberia, but at the present time transport difficulties limit their exploitation. In Chile, Chuquicamata, Poterillos and El Teniente are the major centres. The deposits in Zaire and Zambia make up the so-called 'copper belt' of Africa, in which the deposits of the Katanga district are particularly important. In both countries transport to the coast for export is long and difficult, particularly because of political instability in Angola and Mozambique through whose ports of Lobita and Beira much of the ore is exported.

Bauxite

The importance of bauxite has increased enormously since the Second World War with the growth of demand for aluminium and aluminium products, for which bauxite provides the major raw material. Bauxite, a reddish ore formed by the decomposition of clay rocks in humid tropical conditions, is a hydrated oxide of aluminium. The ore is crushed and concentrated to alumina near the source to minimise transport costs. The alumina is then converted to aluminium in electric furnaces. This final smelting uses vast amounts of electricity and the furnaces are usually located near hydro-electric power stations. Major bauxite producers include Australia, Jamaica, Guyana and France. Both Australia and Jamaica have increased their output enormously in recent years.

Tin

Tin is found both in lode or vein formations in igneous rocks and as an alluvial deposit. It is a soft, weak metal with a low melting point. Its major use is in the tin-plating industry which provides the raw material for food-canning industries, etc. Tin plate is an excellent combination of metals as the steel provides the rigidity while the tin coating provides the crucial resistance to corrosion and ease of soldering. Tin alloys are also used in the engineering industry. Tin is a relatively rare and expensive metal. Major world producers include Malaysia, Bolivia, USSR, Thailand, Indonesia and Australia. Alluvial deposits are typical of Indonesia and Malaysia, whilst Australian and Bolivian tin ores are in the lode form.

Lead

Lead ore or galena is frequently found in association with other minerals including zinc and silver. This association often justifies the recovery of ores which, if they occurred separately, would not be commercially viable. Lead is soft and malleable, making it useful for extrusion, particularly for covering electric cables where its properties of corrosion resistance and poor conduction of electricity provide further advantages. It is also used in various alloys. Major world producers include the USA, USSR, Australia, Canada and Mexico. In the USA lead is mined at Coeur d'Alène in Idaho, Bingham in Utah and Leadville in Colorado. In Australia it is mined in association with silver at Broken Hill and Mount Isa. The Broken Hill deposit is considered to be the largest in the world.

Zinc

Zinc is often found in association with lead as the ore sphalerite. However, it is more costly than lead to reduce and refine from the ore and uses a vast amount of electricity in the process. It is resistant to corrosion and is easily alloyed. Its uses include galvanised sheet, sheet-steel coated with zinc, wire and paint. Major world producers include Canada, the USSR, Australia, USA, Japan and Mexico. Zinc is often smelted and refined in the country of origin, but large quantities of ore concentrated by the flotation process also enter world trade, usually destined for markets in the UK, Germany, Belgium, Japan and the USA.

Nickel

Nickel is used mainly in the manufacture of special steels. Major world producers include Canada, USSR, New Caledonia, and Australia.

Gold

The importance of gold is related to its high value per unit weight and resistance to corrosion which has in the past made it ideal as the basis for national monetary systems and systems of exchange. In recent years this importance has declined as the Gold Standard, the name given to the monetary system based on gold, has given way to standards based on the dollar. Gold is found in two types of deposit, alluvial or placer deposits in which the gold is obtained from streams by panning and dredging, and vein or reef deposits in which the gold is associated with quartz in igneous rocks. The latter deposits are exploited using deep mining methods often involving huge capital expenditure. Major world producers include South Africa, Canada, USA, Japan, Australia, Ghana and the USSR. South Africa accounts for almost 70 per cent of total world production.

Silver
Silver often occurs in association with other metals, particularly lead and zinc, and is frequently recovered as a by-product from the refining of these metals. Major world producers include Canada, Peru, USSR, Mexico, USA and Australia.

Chemical raw materials
A wide variety of minerals are used in the chemical industry, the most important being described below.
Salt is used in the manufacture of glass, soap, caustic soda and bleach. It occurs both in bedded form below ground and as a superficial deposit in desert and semi-desert areas, often as a lake sediment. Major producers include the UK, USSR, France, West Germany and Poland. **Sulphur**, which is used mainly in the production of sulphuric acid, is produced in the USA, USSR, Mexico, France, Finland, China, Spain and Italy. **Potash** is used both as a fertiliser and in the manufacture of explosives, soap, glass and paper. The major world producers are the USA, France, West Germany and East Germany. **Nitrates** are used chiefly as fertilisers and as a raw material in many branches of the chemical industry. Major producers include Chile, Peru and Bolivia.
Phosphates, which are chiefly used as fertilisers, are obtained both from limestone rocks containing phosphate and deposits of guano (bird droppings) found along the coast of Peru. The two leading producers are the USA and USSR.

Key terms
Bedded ore A mineral ore found in sedimentary rock strata.
Flotation process A method of separating valuable minerals from gangue material by utilising the different specific gravities of the two elements.
Gangue The waste material, generally of little or no value, within which valuable minerals occur.
Lode A large mineral-bearing vein.
Opencast mine A large-scale surface mineral working. A quarry. Also referred to as a strip-mine in the USA.
Placer deposit A mineral deposit, especially tin or gold, resulting from the erosion of either bedded or vein deposits and their subsequent redeposition in river gravels or other alluvial material.
Vein A crack or fissure in which minerals have been deposited as a result of igneous or metamorphic action.

Chapter 18
Fuel and Power

All advanced economies are completely dependent on fuel and power for the functioning of industry, transport, etc. It is, therefore, crucial to patterns of employment. This importance is emphasised by the relationship between the consumption of energy and levels of economic development and standards of living; low levels of consumption are typical of underdeveloped nations, high levels typical of the developed countries. This point is illustrated by the following figures for 1972.

Per capita energy consumption in selected countries
(tonnes coal equivalent)

USA	11,611	Turkey	564
USSR	4,767	India	186
UK	5,398	Afghanistan	38

A shortage of fuel and power greatly hinders economic development. However, in developing countries the simple relationship between energy consumption and economic development is distorted, as heavy industry, often characteristic of the early stages of development, consumes more energy than the manufacturing of light consumer goods which is characteristic of the later stages of development. The crucial importance to a country's economy is energy availability rather than cost. Other factors which influence consumption include climate and the possibilities of substitution.

The **energy mix** is the combination of different types of fuel and power which an economy uses. This mix varies from country to country according to factors such as the availability of different sources of fuel, the costs of the different sources, security of supplies, and level of economic development.

Fuel and power resources
The distribution of fuel and power resources is irregular, so that few countries have all sources available to them. Some 90 per cent of coal reserves and 80 per cent of oil are found north of latitude 20°N, while some 65 per cent of water power resources are found south of latitude 20°N. Many factors other than mere presence of a resource determine whether it is exploited or not. The **size** of the reserves is very important in determining their commercial value. Small fuel and power reserves do not justify

148

the large capital outlay needed to develop them. In general the larger a reserve the smaller will be the cost per unit of production. Oil, for example, is found in most countries of the world. The bulk of production however comes from the Persian Gulf, the Caribbean and the Caspian region. Although Middle East oil is a long way from its market, the size of the reserves ensures low production costs per unit and consequently the oil can easily bear the cost of transportation. The nature of coal deposits varies widely from thin, highly-faulted seams to thick, regular, rich seams. Mining activity in the faulted seams is more difficult and consequently more costly with inevitable consequences for the value of the reserve. In the USA the coalfield of Northern Pennsylvania owes its importance to the thickness of its seams and surface outcrops which allow a high level of mechanisation. Similarly, oil is not a uniform commodity; some reserves do not suit the demands of certain markets. Saharan oils, for example, yield far less fuel oil than do Middle East oils and are far more difficult to market in Europe than Middle East oils because of the high demand for fuel oil in Europe.

In the interplay of factors, **access to markets** strongly influences production, as transport can be one of the most costly items in fuel and power production. In Europe, for example, several of the coalfields are badly faulted with thin seams, but because of their proximity to markets are mined economically. On the other hand coal from the USA, where many of the reserves allow cheap, highly mechanised extraction, is able to compete in European markets even after transatlantic shipment.

Labour and **capital** availability also influence which reserves are exploited. Capital becomes increasingly important as the levels of mechanisation increase. This is particularly the case in developing countries, which often have to rely on foreign loans to realise the potential of their resources. The Aswan and Volta schemes in Africa exemplify this point. Different fuel reserves require different amounts of capital for their development. Thus, capital availability can also influence which type of energy source is developed. In Nigeria both hydro-electric power and natural gas resources are found, but because of the lower capital requirements of natural gas, emphasis has been placed on the development of natural gas resources. Sources of fuel and power can be subdivided into primary sources (coal, oil, natural gas, water, geothermal, nuclear and wind) and secondary sources (electricity and coal gas).

Primary sources of fuel and power

Coal

Coal is a sedimentary rock formed by the decomposition and compression of former vegetable matter, chiefly trees, throughout geological time, but particularly in the carboniferous period. Since their deposition many of these coal measures have been highly folded and faulted with consequent effects on the ease of mining. The **nature of the deposits** greatly influences the profitability of mining. In Britain, the more difficult, thin, faulted seams became uneconomic during the 1950s and 1960s when competition from oil closed many collieries mining such deposits. In the 1970s there has been a reassessment of coal as a result of the rises in oil prices, and several thin seams have been reworked. Exploration has also been encouraged and in the UK extensive new deposits have been located near Selby in Yorkshire, and in Oxfordshire.

Several mining methods are used, depending on the nature of the deposits. Where coal occurs near the surface **opencast mines** are used, the topsoil being removed to reveal the coal. This is the cheapest form of mining, as little capital is involved, and is used particularly with the lower grade coals such as lignite. In cases where the coal measures outcrop on hillsides **adit mines** are used in which tunnels are driven along the seams from the outcrop. Where the coal is deep below ground **shaft mines** are used. Deep shafts are sunk down to the coal and galleries excavated to follow the seams. Deep mines are the most expensive, as a considerable amount of capital is involved in sinking the shaft and installing the machinery for cutting and raising the coal to the surface.

Three principal types of coal are mined. **Anthracite** is the hardest form with the highest carbon content. It burns slowly with intense heat. **Bituminous coal** represents the majority of coals, ranging from coking and steam coal, used in power stations, to household coal. **Lignite** has the lowest carbon content and highest water content. Lignite is chiefly used for electricity generation. The chief world producers of coal are tabulated below.

Major coal producers, 1972 (millions of tonnes)

USA	535	UK	119
USSR	451	West Germany	102
China	400	India	74
Poland	150	**World total**	**2,128**

Oil

Oil is a natural hydrocarbon which is formed from decaying organic matter sealed in marine sediments and subsequently trapped between impervious layers of rock. Concentrations are found in favourable geological structures such as fault traps, anticlines, and at the top of salt domes. Exploitation is extremely capital-intensive and is controlled by a small group of major companies, chiefly American. The growth of the oil industry has been very rapid since 1945. Total world production in 1945 was 250 million tonnes, a figure which had risen to 2,000 million tonnes by 1970. The major oil producers are tabulated below. By 1980 it is estimated that the recovery of North Sea oil will have placed the UK among these leading producers.

Major oil producers, 1972 (millions of tonnes)

USA	466	Venezuela	168
USSR	400	Kuwait	151
Saudi Arabia	285	Libya	106
Iran	248	**World total**	**2,401**

As well as being a source of power, oil is also a vital raw material for many industries in most developed economies, forming the basis of the petro-chemical industry which produces a vast range of products from plastics to synthetic rubber. This dependence on oil has created serious economic problems in the developed world following the huge price increases imposed by oil producers since 1974.

The distribution of **oil refining** is controlled by several factors, including **access** to markets, availability of labour, access to deep water, and the availability of extensive areas of suitable flat land. Large economies of scale operate in the refining industry, so that the size of refineries has grown with demand. Refineries with a throughput of 10 million tonnes per year are now common. Because of these economies of scale and the fact that it is easier to transport crude oil rather than refined products, refining tends to take place in oil-consuming rather than oil-producing countries, although countries with small markets may import refined products. Coastal sites dominate the industry, as for example, at Milford Haven, Fawley, Shellhaven, Bordeaux and the Etang de Berre near Marseilles. The growth of large inland markets for oil in Europe has led to the establishment of a number of inland refineries with crude oil being pumped from coastal terminals. Trunk pipelines now run from tanker terminals on the Baltic, North Sea, English Channel, Mediterranean and Adriatic Seas to refineries in West Germany, France, Switzerland, Austria and Italy. Typical of these pipelines is the one-metre-diameter Trans-

Alpine Line (TAL) which runs some 460 kilometres from Trieste on the Adriatic, through the Italian and Austrian Alps to refineries at Ingolstadt in Bavaria.

Natural gas is associated with oil or coal, or is found independently as in the North Sea. Recent discoveries in the North Sea have added the UK to the list of major producers which includes the USA, USSR, Canada, Venezuela, Netherlands and France. Distribution is chiefly direct to market areas by pipeline, although special carriers now allow liquefied gas to be transported by sea.

Secondary sources of fuel and power
Electricity
Electricity may be generated from coal, natural gas, oil, nuclear fuel and water. The choice of fuel depends on many factors, the chief of which are the availability of fuel, the cost of fuel, security of supplies, market size and stage of economic development. The location of power-stations is similarly influenced by many factors depending on the type of fuel used. Three main factors influence the location of **coal-fired** stations; namely, the source of coal, the location of market and the availability of water for cooling. Where low-quality fuel is used, such as lignite in Australia and peat in Eire and the USSR, generation is at the source of raw material. In the UK much of London's electricity is generated along the River Trent in the East Midlands where the combination of factors is particularly economic. However, in France, generation for Paris is at the market, with coal being brought from the Nord coalfield because of the cheap rates available for the movement of coal on French Railways. Other factors which influence the location of coal-fired stations include the availability of labour supply, local subsoil characteristics, ash disposal facilities, and planning and amenity considerations. Similar factors affect the location of gas-fired stations. In the case of **oil-fired** stations there is a tendency for generation to take place near or at the point of refining, thus minimising the transport costs on the oil. In **nuclear** power generation the pull of raw material is completely unimportant since only very small quantities of material are required. Market access consequently becomes an important factor. Because of the possibility of a nuclear accident, planning and amenity issues are also important, so that, although located relatively near the market, nuclear power-stations are found in isolated areas rather than in the urban concentrations themselves. Dungeness power-station serving south-east England exemplifies this point.

Hydro, tidal and geothermal power generation are closely tied to the source of motive power. Both **geothermal** and **tidal** generation are limited to a few areas in the world where opportunities are available, such as Iceland, Italy and New Zealand in the former case, and the Rance Estuary in France in the latter. The potential for **hydro-electric power** (HEP) generation is wider, but because of the huge capital costs of building power stations, most of the world potential is as yet undeveloped. Ideally, rivers with large discharge, rapid flow, and regular régimes, and natural lakes along their courses, provide the best locations, as they reduce the need for expensive dam construction. Many schemes are multi-purpose, being used for irrigation and navigation as well as power production in order to justify the capital expenditure involved. Where flows cannot be regulated, stations with complementary river régimes are often linked. Hydro-electric power-stations in the Italian Alps, for example, where peak flow is in the summer, are linked to stations in the Apennines where the peak occurs in the winter.

Transport of fuel and power

Transport facilities relating to fuel and power can be divided into two groups: **discontinuous forms** such as road, rail, canals and sea, and **continuous forms** such as pipelines and transmission lines. In general, discontinuous forms have low capital costs and high running costs, while continuous forms have high capital costs and low running costs. The variability of transport costs between different modes may be illustrated by the cost of movement of one tonne of oil over 100 kilometres in the USA. Figures are as follows: tanker 6·6–7·9 cents, barge 7·7 cents, pipeline 8·3 cents, rail 45–70 cents, and road $3·50.

Road

The economics of road transport are based on the relatively small capital investment in vehicles, relatively low loading and unloading costs, and high running costs. For small loads and short distances road is the cheapest form of transport, and, as vehicle sizes increase and road improvements are made, it is becoming more competitive over longer distances. A high degree of flexibility is achieved with road transport so that movements can be made easily between mines and factory, or refinery and home. In fact, road transport is frequently the only economic means of final delivery. The use of road transport depends on many factors, but, for example, in coal mining, the age, size and nature of the mine will be important factors. Old mines sunk in the nineteenth century are likely to have sidings already installed giving a strong incentive to use rail rather than road. However, new mines installing a planned distribution system are more likely to use road. The size of the mine and its life-expectancy are important; a large mine will justify the capital cost of sidings. The migratory nature of opencast mining will encourage road transport, while the fixed nature of shaft mining will favour rail. Finally, the nature of the market also influences the mode selected. The large consumer with regular demand and located near a railway will probably be served by rail. The small consumer with irregular demand will be served by road.

Rail

In many countries there has been a serious reassessment in recent years of rail transport for the movement of fuel, as competition from other modes, particularly road, has increased. Rail freight rates are complex, but generally high-value goods cost more to transport than bulky, low-value goods such as coal. Over long distances rail competes favourably for the movement of coal and, to a lesser extent, oil. Over short distances of up to 50 kilometres it cannot compete with road transport.

Water

Sea transport is by far the cheapest form of transport for fuel, especially in the newer bulk-carriers made necessary by the closure of the Suez Canal. The cost per tonne-kilometre in a 100,000 d.wt. tanker is a mere 38 per cent of that in a 16,000 d.wt. tanker. Actual rates fluctuate with demand, and at present are tending to decline with the current surplus in tanker capacity. Transport by sea is very flexible as tankers can be easily diverted to different destinations. Capital costs in sea transport are low, even with the vast size of modern tankers, and make up only 20 per cent of total costs; running costs, made up of labour, fuel and repairs, constitute the significant 80 per cent of costs. With the further development of multi-commodity carriers, costs are likely to fall further as the return journey in ballast is removed. Sea transport involves very high terminal costs, particularly when specialised facilities are required, as, for example, with liquid gas. Terminal costs in the movement of coal from Newcastle to London represent 60 per cent of total transport costs.

Canals have similar advantages for the transport of fuel, and are used extensively in North America and Europe. The main disadvantage relates to the limited flexibility imposed by the available waterway system and the limit to scale economies which the depth of the rivers and waterways bring. However, these limited-scale economies may be overcome to some extent by the use of dracones, flexible plastic bags towed behind barges.

Pipelines

Pipelines are inflexible in terms of routes, capacities and the range of commodities which they carry, but given a steady market, can compete with other land transport forms and occasionally with sea transport. Costs vary with the physical conditions over which the lines pass, the pressure of transmission in the pipes, and the viscosity of the product being pumped. Capital costs are very high, about 75 per cent of the total cost of movement in pipelines. This dominance of capital costs means that unless used at or near full capacity, pipelines are not economic. A further feature of pipelines is the importance of scale economies. As the capital costs of construction do not vary significantly with the size of the pipe, pipelines are particularly favourable for the large-scale movement of commodities on a continuous basis to a regular market. In Europe, with the development of refineries near inland markets, pipelines are used to take crude oil from the coast to the inland refineries. In the UK, where most refineries are on the coast, pipelines are chiefly used for oil products rather than crude oil, and only where deep water is not available near refineries are pipelines used for crude oil, as, for example, between Milford Haven and Swansea, and Tranmere and Stanlow.

Transmission lines

Transmission lines have basically the same economies as pipelines: high capital costs, low running costs, and the need for a large regular load. Costs of transmission are uniform up to about 700 kilometres, after which leakage of power increases unit costs. Distances of transmission have been increased in recent years by using direct rather than alternating current. Underwater cables link Sweden and Denmark, Britain and France and the islands of New Zealand.

Key terms

Adit mine A type of mine in which a more or less horizontal gallery is driven into the ground, usually from a hillside.
Anthracite A hard, high-grade coal with over 85 per cent carbon content.
Bituminous coal Coal of medium quality including coking coal, steam coal and household coal. 75–85 per cent carbon content.
Energy mix The combination of different types of fuel and power which a national economy uses.

Chapter 19
Industry

The term 'industry' encompasses a wide range of activities generally undertaken for profit. It includes **primary industry** such as fishing, forestry, agriculture and mining, **secondary industry** or manufacturing which is concerned with conversion of raw materials into new products, and **tertiary industry** or service industry, which provides services such as insurance, banking, research and other professional work, often for manufacturing industry. In terms of employment, both the primary and secondary sectors have declined since the war, although their contribution to economic growth is still of crucial importance. This chapter deals with manufacturing. Within the manufacturing sector it is possible to distinguish two basic types; namely, the production of capital or producer goods, and the production of consumer goods. **Capital goods** are required for the production of other goods; examples include machine tools and processing equipment. **Consumer goods**, as the name implies, are goods produced for final demand, for sale to consumers. The complexity of modern manufacturing industry makes this distinction less clear, as many products are neither capital nor consumer goods but form the raw materials, in the form of components, for either.

The location of industry

Each type of manufacture has different requirements, but the patterns of world manufacturing allow general statements to be made about the factors which influence the location of industry. In some cases all factors are important, in other cases very few. It is therefore important to appreciate that the factors discussed below do not necessarily apply to all locations and industries.

Fuel and raw materials

In the early stages of industrial development, before the canal and railway eras, industry was very closely linked to sources of raw materials. The early UK iron industry illustrates this reliance, with the iron-makers establishing furnaces close to sources of iron ore and charcoal in the Weald and Forest of Dean. Although still important, particularly where large quantities of low-value bulky materials are required, the importance of raw materials as a locational factor is declining. Improved transport has allowed materials to be moved more cheaply, thus allowing other factors to control location. Some industries are still linked to materials and power; these include brick and cement manufacture.

Transport

Many aspects of transport are important to industrial location. Without transport facilities raw materials cannot be assembled or products distributed. The absence of transport facilities still limits manufacturing in many underdeveloped countries. The cost and modes of transport are important to location. Cost, for example, determines the distance that production can take place away from either the source of raw materials or market. Early theories of industrial location were largely based on the cost of transport, and suggested that production will take place where the costs of assembling materials are at a minimum. More recently it has been suggested that since transport costs are only a small proportion of total production costs in many manufacturing processes, they are of little significance in all but a few industries using heavy, bulky raw materials and producing low-value, bulky products.

Labour-supply

Although industry is becoming increasingly mechanised, the availability of labour is still an important locational factor. The existence of a labour force of sufficient **size** is important, but the particular **skills** of that labour force are often more significant. Industry is attracted to areas where the right type of labour is available. Pools of labour with particular skills develop through time; Sheffield has skilled steel-workers, Glasgow skilled ship-builders, Birmingham craftsmen gunsmiths, etc. The **cost** of labour as a proportion of total production costs varies from industry to industry. In the USA, for example, labour contributes 60 per cent of total value added in the leather industry, but only 29 per cent in the chemical industry. Thus, cost of labour will be a more important factor in the choice of sites for the leather industry than the chemical industry. Labour costs also vary geographically. This variation is caused more by variations in productivity than actual wage rates, because wage bargaining has made national basic rates common in most countries. Male and female labour is required in different quantities by different industries. The availability of either male or female labour may influence the choice of location for particular industries.

Capital

Capital can be subdivided into two elements: goods and equipment, often referred to as **fixed capital**, and **money capital**. As the name implies, fixed capital, including buildings and machines, is immobile, and tends to keep industry in the same place until its useful life has been completed or it has been written off. This leads to geographical inertia (see page 160). Money capital on the other hand is highly mobile, as demonstrated by the involvement of the London and New York capital markets in world-wide industrial developments.

Market demand

Without a market no production can take place profitably. The importance of markets is linked to transport as well as the specific characteristics of the market itself. **Access** to market is crucial. If it cannot be reached it might as well not exist. The market factor is particularly important to industries which have no specific locational requirements, especially those concerned with the manufacture of consumer goods. Of the specific market characteristics which are important, market size and spending power or affluence are probably the most important. Market **size** influences ease of distribution, a large concentrated market being easier to service than a small dispersed one. As industrial processes become more complex and the size of production units increases to take advantage of scale economies, access to a large concentrated market is becoming more and more important. This trend may be observed in many countries. In the UK industrial growth in recent decades has been most marked in the South East with the large London market for goods. Market **affluence** affects location in a similar way to market size, as the potential sales of a particular manufacturer will be controlled by the purchasing power as well as the number of potential purchasers.

Government influence

It is not only in communist countries that governments influence industrial location. In the western world governments have both a direct and indirect influence upon industry. Direct influences are principally in the form of relocation policies which attempt to direct industry to areas of high unemployment. These policies operate in most Western countries. In the UK relocation policies have influenced location, with varying degrees of success since the Special Areas Act of 1934. At present aid is given to firms wanting to relocate in the Special Development Areas, Development Areas and Intermediate Areas, and Northern Ireland. This aid takes the form of tax allowances, grants, loans and subsidised factories. The designated areas cover a large part of the country, including the South West, Wales, Merseyside, much of Yorkshire, parts of Derbyshire, much of northern England and Scotland. Indirect influences are many, and include ownership of companies such as the British Steel Corporation and the National Coal Board, freight pricing policies of British Rail and British Road Services, and public investment in motorways which improve regional accessibility. These direct and indirect influences are both positive in that they attempt to attract industry to particular regions. However, there are other negative government influences which attempt to prevent industry moving to a particular place or region. These negative controls include both planning regulations and the industrial-development and office-development certificate system which requires firms to apply for a certificate for any large new development. These certificates are not issued where developments are regarded as undesirable.

Industrial concentration

Industry may be attracted to an area by energy, raw materials, labour-supply or some other specific factor. As an area develops, other industries will tend to be attracted to it. These may use the products which were previously wasted, or new firms may be established to supply goods to the older firms. The area will consequently grow in industrial importance and with the growth of labour a large consumer market will develop which in turn will attract light consumer industries. The concentration of firms which develops in this way, and the links which develop between firms, bring external economies to the individual firm. These external economies can be made by an individual firm sub-contracting work out, as in motor manufacturing, where many components are made by small specialist firms, and by the availability of banking, repair and maintenance and research facilities. The importance of such external economies has increased since the war and has led to a further concentration of manufacturing activity. However, there is a maximum economic size for both industrial plant and industrial groupings, so that at some point there will be no further advantage in concentration. There is some evidence that this position has been reached in South East England where the costs of land, labour and distribution have reached very high levels, so that many firms are now tending to locate away from the concentration. External economies derived from concentration have given rise to the major manufacturing regions of the world. In Europe a belt of industrial concentrations includes north-eastern France, the Sambre–Meuse Depression, Kempenland, South Limburg, the Saar, Aachen and the Ruhr.

Geographical inertia

Geographical inertia is the term given to the situation whereby industry remains in the same location although the initial reasons for locating there have long disappeared. Several factors explain this situation. As discussed earlier, capital equipment is immobile and often has a very long life so that it is not easily written off. A firm will prefer to continue using that equipment at the original location than invest in new capital at a new location. Furthermore, it is often cheaper to expand capacity on an existing site than to establish a new one. It is estimated to be 300 per cent more expensive to build a new steel works than to expand an old one to an equivalent capacity. The old area will have the attributes associated with external economies such as transport facilities and pools of skilled labour. Geographical inertia is perhaps most easily observed in heavy industrial areas such as Pittsburgh and Sheffield.

Other factors

Many other factors affect the location of industry. These include the availability of flat land for factories, the load-bearing character of the land, water supply, ease of effluent disposal, etc. Increasing concern has also been shown in recent years regarding the effect of industry on the quality of the environment. Planning permission for industry is now becoming increasingly difficult in areas where it is likely to create an intrusion in the landscape.

The major manufacturing industries

Iron and steel

Early developments in the iron and steel industry can be traced back to the Middle Ages when deposits of iron ore were quarried and smelted by means of charcoal. The industry at that time was very small in scale and closely tied to its sources of raw materials, iron ore and charcoal. The industry moved as the raw materials were exhausted, but the principal locations in the UK were the Weald and the Forest of Dean. The beginnings of the modern industry came in 1705 when Abraham Darby discovered a method of smelting iron using coal at Coalbrookdale in Shropshire. From that time the locational ties with the coalfields developed. The basis of modern iron-making is the blast furnace in which a mixture of iron ore, coke and limestone is blasted by hot air. As the industry has developed, furnaces have been increased both in size and efficiency. The average output per furnace in 1788 was 912 tonnes per annum. In 1970 Nippon Steel's Nagoya Works No. 3 furnace produced 10,080 tonnes of iron in twenty-four hours. A second major development has been a reduction in the amount of coke required in the blast furnace. In

1828 James Neilson developed the use of a heated blast, which reduced the coke input from 8 to 5 tonnes per tonne of iron. Large modern furnaces have further reduced this coke requirement by their internal efficiency.

Steel is essentially iron with the addition of carbon and other metals. The carbon content ranges from 0·3 per cent to 1·5 per cent. Other metals added for special steels include nickel, chromium, molybdenum and tungsten. A revolution in steelmaking came in the 1850s with the invention of the Bessemer Converter and the Siemens-Martin open-hearth furnace. Up to that time steel had been very expensive and used only for the manufacture of swords and springs. The Gilchrist-Thomas process allowed the use of phosphoric iron ores, following which the steel industry expanded rapidly in many countries. The more recently developed electric furnace has allowed steel-making to be more controlled and is important in the making of special steels where small quantities of high-grade steel are produced.

The main raw materials in iron- and steel-making are coke, iron ore, limestone, and water. Because of their bulk and high cost of transport the location of the iron and steel industry has traditionally been on coalfields or iron-ore deposits or at a break-of-bulk point such as a coastal location. The steel works at Dunkirk in France and most Japanese works are good examples of coastal locations, particularly in the Japanese case where all the raw materials are imported by sea. Technical improvements in manufacture have reduced the amount of coke per tonne of iron produced, and reduced the pull of the coalfields. The invention of the electric furnace completely removed the locational attraction of coal, although because of industrial inertia, many centres of special steel manufacture are still located on or near coalfields. In special steel manufacture scrap iron and steel are often used instead of pig iron and the nature of the processes encourages location where pools of skilled labour are available.

Major steel-producing areas
USSR The iron and steel industry is well developed in the Donbas, Donetz, Krivoirog, Kuzbas, Urals, Caucasia, Irkhutsk and Leningrad areas. The industry is closely related to the planned development of manufacturing regions established throughout the Soviet Union. Great dependence is placed on the movement of raw materials by rail using return-haul methods. Total production is 120 million tonnes per year.
USA The major centres of iron and steel production are

Pittsburgh, the Mid-Atlantic seaboard, and various towns such as Gary and Buffalo on the shore of the Great Lakes. All are old-established centres and make considerable use of cheap water transport and return hauls of either iron ore or coal. Birmingham is another major centre in the south where local coal and iron ore deposits and cheaper labour are used to supply the southern markets. More recent developments have taken place in Los Angeles to supply the growing West Coast markets. Total production is 120 million tonnes.

Japan The rise of Japan as a major steel producer has been dramatic. Production rose from 5 million tonnes in 1950 to 103 million tonnes in 1970. This rise was associated with the rapid post-war growth of industry in the country which produced a rapid growth in demand for steel. Japan now boasts the largest and most modern steel works in the world. Depending as she does on imported raw materials, works are mostly located on coastal sites as at Tokyo, Kawasaki, Kobe, Osaka and Yawata.

West Germany The principal area is the Ruhr, particularly the towns of Dortmund and Essen. Use is made of both local and imported coal and iron ore imported via the Rhine. Production is 43 million tonnes per year.

United Kingdom Although most production is located in South Wales and the North East, the British Steel Corporation's modernisation plans will further concentrate production into much larger units, particularly in the north-east. Smaller works such as those at Shotton in North Wales and Ebbw Vale in South Wales will be closed. Total production is 25 million tonnes per year.

Australia The major centres in Australia are located on or near the coast, making use of cheap water transport and return hauls. Major works are found at Newcastle, Port Kembla, Whyalla and Kwinana. Total production is 6·5 million tonnes per year.

Engineering

A variety of industries make up the engineering industry, from heavy engineering, such as shipbuilding and structural engineering, to light engineering, which is concerned with the production of radios and other light electrical goods, etc. Although basically tied to iron and steel as the main raw materials, the locational patterns of each type differ. In the case of light engineering the market attraction is often particularly important.

Shipbuilding The chief world shipbuilding nations and the gross tonnage produced by each in 1972 are as follows: Japan, 12·8 million tonnes; Sweden, 1·8 million tonnes; West Germany, 1·6 million tonnes; UK, 1·2 million tonnes; Spain, 1·1 million tonnes; France, 1·1 million tonnes; USA, 0·6 million tonnes. The major locational requirements of access to steel supply and deep water for launching are reflected in the location of the industry; for example, at Göteborg in Sweden, and along the estuaries of the Clyde, Tyne, Wear, Tees and Mersey in the UK. The relative importance of the UK industry has declined as competition from other nations, particularly Japan, has grown.

Motor industry The motor industry is extremely complex, including as it does hundreds of small firms producing individual components, as well as the major manufacturers who in many cases only assemble these components. External economies are very important to the industry. However, location is often influenced by historical legacy as in the case of the Austin/Morris works at Cowley. Where new locations are sought, market-oriented sites are popular, such as that of Ford at Dagenham. The motor industry has grown considerably since the Second World War and has consequently been at the forefront of government attempts to divert industry to the less prosperous regions. In the UK British Leyland has plants in Scotland, Cumbria and South Wales, while Ford, General Motors and Chrysler have plants on Merseyside. All are a result of government intervention.

Radio and electronics Radio and electronics have witnessed a rapid growth since 1945, particularly with the invention and mass production of transistors and the development of computers. Within the industry, both large international firms and small subcontracting companies are found. Again, external economies are important, and, as with motor manufacture, the large companies often only assemble products. Two locational factors are important. The consumer orientation of the industry draws it to the market, and the constant technological improvement and research involved draws it towards research establishments where both new ideas and research personnel are available. This pattern is exemplified by the concentration of the industry around Boston in New England. Radio and electronics is a growth industry and has been the subject of government intervention in many countries. Many large companies, especially those employing a high proportion of female labour, have been established in the less prosperous regions of many economically advanced countries.

Chemicals

Although extremely diverse in terms of its products, the chemical industry is made up of a few very large companies. Early development of the industry was based on salt, and was consequently located near salt deposits, as, for example, in Cheshire and the North East in the UK. Later development of the industry was based on coal and more recently on oil, and this has led to locations on coalfields or near oil refineries, although geographical inertia still keeps the industry in many of the traditional areas. Major producers of chemicals include the USA, USSR, West Germany, France and the UK.

Textiles

At present the production of textiles is one of the world's leading industrial activities and has a world-wide distribution. However, the textile industry's early development in the UK and other West European countries was small-scale and based on local supplies of wool and the availability of soft water. In the Middle Ages, East Anglia, Oxfordshire, and the south-west were important centres in the UK. With the advent of mechanisation, and particularly the use of steam power, the industry developed in Yorkshire, not only because of the availability of local wool and coal, but also because other areas were slow to adopt the new technology. The West Riding still maintains its world position for the production of fine woollen textiles. Major world producers of wool textiles include the USSR, UK, Japan, Italy, France, USA and Belgium.

In its early stages the UK cotton industry was located at several centres on the west coast, where water, damp atmosphere and power supplies were available and where cotton could be easily imported. As their economy grew, some of these areas developed specialisms in other industries, notably shipbuilding on the Clyde, leaving Lancashire as the main centre for cotton manufacture. In recent years the UK cotton industry has declined in the face of competition from abroad, particularly from India, Hong Kong, and Taiwan. Leading producers of cotton textiles include the USA, USSR, China, India, Japan, Pakistan and France.

Manmade fibres include rayon, nylon and terylene. The manufacture of these fibres is closely related to the chemical industry. The weaving of the fibres tends to adopt similar locational patterns to those of the weaving of natural fibres.

Industry and underdeveloped countries

Most underdeveloped countries regard industrialisation as the means of raising living standards and stimulating their economic development. However, many factors make the large-scale development of industry difficult in such countries. These factors include problems of labour supply, lack of capital, and marketing difficulties. These problems will be examined in turn. On page 157 the importance of labour-supply was discussed, and this is particularly important in the developing world. Most industrial processes require some levels of skilled and semi-skilled labour. The factory system also imposes a level of discipline on the work force to which it may not be accustomed. In many countries these labour requirements cannot be met, and it may be necessary to recruit skilled workers from other countries. Low productivity and high labour turnover are common in developing countries.

Unless endowed with valuable raw materials for export, as in the case of the Middle East oil producers, shortage of capital is a crucial problem to most underdeveloped countries, which need to import capital goods and machinery to establish and operate their factories. External sources of capital are available either from the developed nations of the world or agencies such as the World Bank. However, such loans are burdensome to developing economies and may have unacceptable political conditions attached.

As mentioned earlier a market for products is necessary for any industrial activity to be successful. Competition with old-established producers in export markets is difficult. Thus, a large home market is vital for any developing country. Unfortunately, such countries generally have a dispersed population with low purchasing power, thus making distribution and marketing of products very difficult.

Given these conditions, early industrialisation in developing countries is most successful where it concentrates on those products which have a large local market, require relatively little capital, and make use of unskilled labour. Textile manufacture fulfils many of these requirements, and is characteristic of the early stages of industrialisation. It provides a foundation on which the necessary economic and social changes can be made, and on which other industries can build.

Key terms

Capital goods Goods, such as machine tools, required for the production of other goods.
Consumer goods Goods, such as furniture, produced for final demand, for sale to consumers.
Geographical inertia The tendency of an industry to remain in a given location long after the conditions originally determining that location have ceased to operate or exert any influence.

Chapter 20
Settlement

Rural settlement

In the classification of rural settlement patterns a basic twofold distinction can be made between dispersed and nucleated settlements. A **dispersed settlement** pattern is one in which the unit of settlement is the individual farm or homestead, and these are spread at some distance apart throughout the parish. Such a situation exists where farms are set in the centre of their fields or spread out along roads rather than being concentrated at one point, and is common in many sparsely populated upland regions and in areas where settlement has occurred comparatively recently following land drainage or reclamation. For example, much settlement in the East Anglian Fens is dispersed along the roads and drainage channels, although the population density is relatively high.

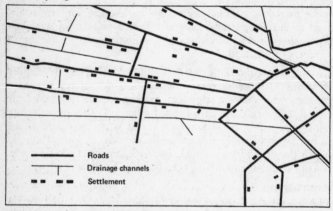

Figure 44. Dispersed settlement in the Fens

Nucleated settlements are those where houses and other buildings are grouped around some central feature, frequently a church, green or crossroads, with the exception of perhaps a few isolated farms. Such nucleated settlements are termed hamlets or villages according to their size and function. A **hamlet** is a small group of some ten to twenty homesteads and farms, but usually lacks most of the basic service facilities. A **village**, however, is a larger nucleus and is thus able to support services such as a sub-post office, general store, church and community centre. Even higher up the scale are large **urban villages** and small **market towns** which can provide a larger number and wider range of services. Although the distinction between dispersed settlement, hamlets, villages and towns is a useful one, it must be remembered that in many areas the settlements do not fall conveniently into any one category, since there are numerous intermediate situations.

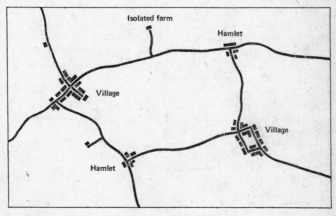

Figure 45. Nucleated settlement

The density and form which any rural settlement takes is closely influenced by physical factors. Areas of fertile soils, land free from flooding, south-facing slopes and with good water supply are the kinds of sites likely to attract and support a dense population. Upland regions, poor soils and land liable to be flooded are unattractive to settlement. However, it is also important to take into account historical, economic and social factors such as the need for defence, the history of colonisation, traditions of society and land use, and the development of transport facilities.

A number of characteristic village types can be identified on the basis of their site, function and form (see figure 46).

Spring-line villages have developed on sites where advantage could be taken of springs as a source of water supply. Frequently

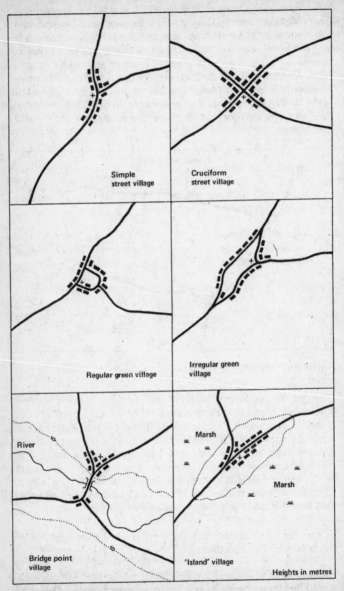

Figure 46. Some village types

168

such villages are grouped along a line of springs at the foot of a scarp slope where two contrasting rock types meet, as for example along the foot of the Chilterns and Cotswolds escarpments.

Valley flood-plain villages are usually sited on dry points away from flood risk, commonly along river terraces or where the land begins to rise at the edge of the flood-plain. The water supply provided by the river, fertile alluvial soils and the use of the valley as a line of communication may all be contributory factors favouring this kind of site.

Hill-top villages normally fall into two categories. Sometimes the site was chosen to avoid flooding in marshy areas or on flood-plains, such as Haddenham and Sutton in the fenland, but more commonly the reason was a concern with defence.

Bridge-point villages, sited at important river crossings, also had a defensive significance and often developed as centres of communications owing to the convergence of roads on the bridge.

Street villages are settlements which have grown up in a linear form along a road. The simplest form is based on a single road but variations occur based on the existence of a parallel back street, a road junction (T-shaped village), or a crossroads (cruciform village).

Green villages are one of the commonest British village types, with the buildings arranged around a central green. Again there is a variety of forms based on different shapes of green, such as the square, triangular, elongated and irregular. The nature of the green as a focal point of the village is often emphasised by the existence on it of the church, pump or pond.

Urban settlement
Functional classification of towns
One of the most important ways of distinguishing between villages and towns is by the services which they provide and the functions they perform. Whereas villages offer only a few basic services for their inhabitants and those of nearby farms and hamlets, towns provide a much wider range of services and act as shopping and market centres for the surrounding rural areas. The larger towns and cities provide even more specialised services for hinterlands which may include rural areas and smaller towns, and they contain the larger shops, department stores and the more

important banking and professional services. An excellent example of a city which is basically a major market centre is Norwich, which provides for, and is supported by, the surrounding agricultural districts of Norfolk. However, each town performs many other functions and one way of classifying urban settlements is to identify their dominant function.

Commercial centres are towns concerned mainly with finance and trade, having specialised banking and insurance facilities, such as Zürich and Frankfurt.

Mining towns are often located in unusual or unfavourable places owing to their dependence on the exploitation of mineral resources. The city of Johannesburg is based on gold mining, and Kiruna (Sweden) owes its existence to the mining of iron ore.

Industrial towns are centres in which the dominant activity is the processing of raw materials. Many towns in this category have a wide range of different industries, but others may be dominated by a single industry such as Sheffield and Pittsburgh (steel).

Administrative centres Many cities are dominated by a function as the capital of a county, state or country. Examples include Canberra (Australia), Brasilia (Brazil) and Washington D.C. (USA). In these cities a high proportion of the working population is engaged in administration.

Cultural and educational centres The existence of an ancient and world-famous university dominates towns such as Oxford, Cambridge, Heidelberg (West Germany) and Lund (Sweden).

Religious centres A few towns have achieved a special status as the objects of religious pilgrimages, notably Jerusalem, Mecca and Lourdes.

Tourist centres and holiday resorts Favourable surroundings or a special characteristic such as a spa may enable towns to develop as tourist resorts. Many are coastal towns, such as Blackpool and Miami, but there are also a number of inland resorts such as Bath and Innsbruck.

Residential towns Around many large cities are to be found a few **'dormitory' towns** which serve mainly as the home of workers who travel into the cities each day. Examples include Dorking (London) and Altrincham (Manchester).

New towns have been developed in order to relieve congestion in the large cities by acting as counter-attractions. In Britain examples include Crawley, Redditch and Cumbernauld.

Transport and route centres Certain towns have become dominated by a transport function, owing to a strategic location. Crewe and Chicago are both important railway centres, for example. The most significant group of transport towns are, however, the **ports**, which can be subdivided into several different types. The **commercial ports**, such as Glasgow, Southampton and Marseilles, have a wide range of facilities to cater for general cargo and passenger traffic. **Entrepôts** specialise in the redistribution of cargo, the best examples being Rotterdam and Singapore. **Packet stations** such as Dover, Calais and Ostend deal with short-distance ferry traffic, whilst Hull and Grimsby are essentially **fishing ports**. Other types include **tanker ports** (Milford Haven), **outports** (Avonmouth), **ports of call** (Honolulu) and **naval ports** (Chatham).

Town sites
Another method of classifying towns may be made on the basis of their site and position, although it should be remembered that a dominant factor rather than a single cause for their growth is under discussion.

River sites Rivers have exerted a very significant influence on the siting of many urban settlements, the following being examples of some of the major types: **river gap** (Lincoln, Toulouse), **meander core** (Durham, Shrewsbury), **bridging point** (Bedford, Montreal), **confluence** (Lyons, Khartoum), **delta** (Cairo, Calcutta), at the **end of a gorge section** (Bonn, Vienna) and at the **head of navigation** (Rouen, Hamburg).

Coastal sites The coastal plains between the sea and uplands offer an attraction for urban development, in addition to the more obvious functional categories of ports and seaside resorts.

Lakeside sites are often favoured as transport centres, as well as being attractive tourist resorts; for example, Geneva and Como.

Fertile alluvial plains Many important towns are sited at the centre of fertile lowlands, Paris and Norwich for example, whilst another common location is at the junction of a lowland plain and

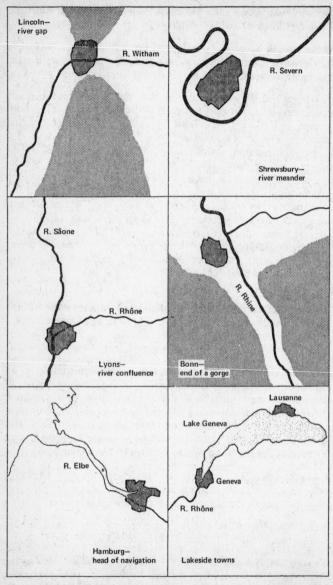

Figure 47. Some town sites

an upland region, as at Denver and Turin. Both these sites are likely also to be the focus of routeways.

Other types of town sites include **upland plateaux** (Mexico City), **marsh 'islands'** (Ely), **defensive locations** (Edinburgh) and **frontier towns** (Basle).

Urban structure

Within towns and cities certain distinctive areas or structural zones may be recognised, reflecting different types of land use and stages of growth. The central area of a town, or **central business district** (CBD), is the zone in which are concentrated the main shopping facilities, financial and professional services, offices and administrative functions. The CBD is characterised by high property values and rents, and thus maximum use of space is desirable, a situation which is often achieved nowadays by building upwards. For this reason skyscraper office-blocks have become common features of the central areas of towns.

Surrounding the central area is a **'blighted' zone**, a district in which poor-quality housing is intermixed with industry and small businesses unable to afford the cost of a central site. Printing, clothing, motor-vehicle repairs and 'backyard' workshops are the kinds of enterprise characteristic of this area.

The main **industrial zones** are usually to be found around the major transport links, close to the railways, rivers, canals and docks. However, modern light industry is generally located on the edge of the urban area, particularly where the access to main roads and motorways is good. In many cases specially designed **industrial estates** have been built for this purpose.

There are several different types of **residential districts** within towns, and these are distinguished by differing housing densities and by their characteristic building styles and street patterns. Typical nineteenth-century housing is fairly dense, built on regular street plans in terraces of two or three storeys, and normally forms an inner residential zone. Much of this housing is of poor quality and is now being replaced by new buildings which, in order to retain the high housing density, are often multi-storey blocks. During the twentieth century towns have undergone very rapid expansion by means of **ribbon development** along main routes and by the construction of vast new estates on the outskirts. The early estates can easily be recognised since they were built to regular geometric patterns, whilst in more recent years informal, irregular plans have been preferred.

Various **open spaces** exist within any urban area, and the basic distinction to be made is between developed open space and derelict land. The former category includes facilities such as parks, playing fields, recreation grounds and botanic gardens, whilst the latter is often associated with land which is physically not suitable for building development owing to such factors as the risk of flooding, steep slopes or subsidence.

Problems of urban growth

The rapid growth of towns and cities in recent years has brought with it a number of problems. In developed countries such as the UK the majority of the population lives in urban areas, a trend which has resulted partly from the decline in agricultural employment in the countryside so that people have been attracted by the prospect of jobs and the 'city life'. In many cases this has put pressure on the provision of various services in towns, particularly housing, with the consequence that there is frequent overcrowding and a fall in living standards. Such a deterioration is usually clearest in the inner urban areas (hence the 'blighted' zone) and has meant that many people are now moving to the outer suburbs or rural fringes, and the inner areas are suffering an actual loss of population.

Another problem is that of urban transport, for people need to travel between the residential areas and other parts of the town to work, shop and for recreation. In most large towns the roads are congested and public transport overloaded during the morning and evening rush-hours, a problem made worse by the increase in population in the outer areas.

In the underdeveloped countries the problem of urban growth and a rapidly deteriorating living standard is even greater. People have been attracted to cities only to find both jobs and housing scarce, with the result that many thousands are forced to live in slums and **shanty towns**.

Key terms

Central business district The central area of a large town. The city district dominated by commercial and administrative functions. Often referred to simply as the CBD of a city. In the case of American cities, also known as the downtown district.

Cruciform village A village which has grown up at the inter-section of two routes, with houses and other buildings along each axis.

Dispersed settlement A pattern of rural settlement in which most people live in isolated houses or cottages so that clusters of settlement such as hamlets and villages tend to be absent.

Dormitory town A town dominated by residential areas and often lacking in many urban functions. Usually found around a larger central city. The home of commuters who travel into the central city each day.

Entrepôt A type of commercial port concerned especially with the trans-shipment or redistribution of cargoes.

Green village A type of village with buildings arranged around a central green or open space.

Industrial estate A planned agglomeration of factories and workshops. A district of light industry, especially concerned with the production of consumer goods. Usually located close to the city margin.

New town A relatively recent, planned urban settlement, usually designed to attract population from another congested urban centre. Sometimes referred to as a satellite town or overspill town.

Nucleated settlement A pattern of settlement in which houses and other buildings are clustered closely together in hamlet and village forms.

Outport A port located downstream from another port. Usually developed as a result of an earlier port becoming silted or inaccessible to large modern vessels.

Packet station A small port dealing with short-distance ferry traffic.

Ribbon development A linear development of houses along a main road, extending outwards from a town. A typical form of inter-war suburban development.

Spring-line village A village which has grown up around a spring which provided an early source of water supply.

Street village A village which has grown up along a road, and which is usually linear in form.

Chapter 21
Transport

The significance of transport

One of the bases of economic activity is that of trade between different areas producing different goods, and transport is of vital importance as the means by which this trade takes place. In many ways the process of transfer may be seen as an integral part of the production process, and for this reason there is a close link between the transport network of a region and its level of economic development.

The type and quality of the transport system in any area varies according to various physical, economic and historical factors. The earliest forms of transport were human porterage, pack animals and draught animals, and whilst these have been superseded in most parts of the world they are still used in some remote places where physical conditions are severe and roads poor or non-existent. For example, llamas are an important means of transport in the high Andes, yaks in Tibet and huskies in the Arctic. Metalled roads were built in Roman times, but after the fall of the Roman Empire the art was lost and did not reappear until the eighteenth century, when great engineers such as Telford and Macadam were active in Britain. This period saw the beginning of what may be called the 'Transport Revolution', for the eighteenth century was the main period of canal construction. More important still was the advent of the railway in the early nineteenth century, which for the first time provided a transport system that was both fast and cheap, and which played a major part in the spectacular economic development of Victorian Britain. In the twentieth century the development of motor vehicles transformed road transport and has led to the decline of the railways, whilst the invention of the aeroplane (in particular the jet aeroplane) may justifiably be said to have brought about another revolution in transport.

Inland transport

Roads

Ranging in nature from rough, unmetalled tracks to modern, multi-laned highways costing many millions of pounds to con-

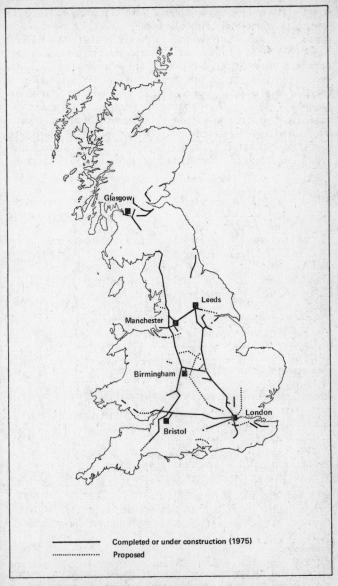

Completed or under construction (1975)
.... Proposed

Figure 48. Motorways in Great Britain, 1975

struct, roads are the most universal form of transport. There are few places where they cannot be built since, unlike railways and canals, they are relatively free from the restrictions of relief, although adverse weather conditions (particularly snow and fog) may be a disadvantage. Since roads are so widespread and link even the smallest settlements, the various forms of road transport are able to provide a door-to-door service, which the railways cannot do without trans-shipment. Thus, road transport has tended to replace the railways for the movement of passengers and freight over short distances, and is competing strongly for much of the long-distance traffic for which the railways are better suited. At present 64 per cent of freight movement in Britain is by road.

Most countries in Europe and North America have well-developed, dense road networks, but elsewhere few exist outside the main urban areas. New long-distance roads are being constructed to help open up regions of the world where transport is poor and is hindering economic development: for example, the Pan-American Highway will eventually link North, Central and South America, and the Stuart Highway gives access to the remote outback of central and northern Australia. In Britain the rapid growth in the number of vehicles on the roads (which is expected to double by the year 2000) means that the road network must continually be adapted and modified to cope with congestion. The most important way in which this is being done is by the construction of **motorways**, which are to form a completely new system of arterial roads designed for the use of fast motor traffic only and with limited access from the pre-existing roads. Already over 2,500 kilometres of motorways have been opened, and over 3,200 kilometres are planned for completion by the early 1980s (see figure 48). Similar new road systems are being built in most other developed countries, for example the German *Autobahnen* and Italian *autostrade*. Another significant improvement in the British road network in recent years has been the construction of several major estuarial **bridges**, which make the avoidance of considerable detours or ferry crossings possible. Examples include the Severn, Forth and Tay bridges, and one is at present being built over the Humber.

Railways

Railway networks as dense as those of roads are not possible owing to the much higher costs of railway construction, and technical limits (slight gradients and gentle curves) which put a restriction on the choice of possible railway routes. Nevertheless, the railways played a vital part in nineteenth-century economic development, when road transport was far less efficient, and had an important role in opening up new lands where other forms of transport were inadequate – notably such lines as the Canadian Pacific Railway and the Trans-Siberian Railway. In this context railways are still being built in many developing countries, particularly in Africa.

In Britain and other developed countries the pattern of railway development has exerted an important influence on the location of industry and on urban development, and thus the railway network in an area is frequently a reflection of its economic geography. However, a characteristic of many railway systems is that they were not developed as national systems but were built by private companies, and this has often given rise to problems of interconnection between lines and the duplication of routes. Most railways are of the standard gauge of 143 centimetres, but some lines are broad- or narrow-gauge, the latter being useful in mountainous areas since it enables sharper curves to be negotiated. In Australia, the various states built railways to differing gauges, and the resulting problems of traffic interchange are only now being overcome, by converting more lines to standard gauge. Motive power is mainly by means of diesel engines or electricity, but steam locomotives are still of importance in a few countries.

Although the railway network is still being expanded in some parts of the world, notably Japan and the developing countries, the competition of road transport has resulted in the contraction of most other systems. In Britain the route mileage has been reduced by nearly 11,000 kilometres to about 18,500 kilometres since 1961, and only 18 per cent of freight is now carried by rail. Much of this is long-distance transport of bulky goods, to which the railways are most suited, and special bulk trains and container trains (**freightliners**) have been designed for this traffic.

Inland waterways
Although inland waterways are a very cheap form of transport they are also extremely slow, and are therefore best suited to the movement of bulky, low-value goods. Few rivers can be used as waterways without modification, owing to variations in water level and frequency of meanders, and therefore in most cases artificial cuts and canals are built to provide a more flexible route.

In Britain the canal network is no longer of much importance, with the exception of a few relatively short routes to inland ports, such as the Manchester Ship Canal and the Gloucester and Sharpness Canal. In Europe, however, numerous canals link the major rivers such as the Rhine, Seine and Danube to form an important, fairly dense network on the North European Plain. In North America the Great Lakes and the St Lawrence Seaway together form one of the world's most important inland waterways, whilst the Panama and Suez Canals are of vital strategic importance in linking the oceans on two sides of an isthmus.

Pipelines

In recent years pipelines have become a very significant element in inland transport, particularly in connection with the movement of oil, oil products and natural gas. The cost of laying pipelines is very high indeed, but they are the cheapest method of carrying large quantities of liquid or gas regularly over one particular route.

Sea transport

Transport by sea remains very slow, but ocean shipping is one of the cheapest means of moving goods and passengers since the route is a natural one, and therefore no construction costs are involved. Nevertheless, trade passes along certain well-defined ocean routes, and it is necessary to provide terminal facilities for loading and unloading.

The busiest and most important shipping route is the North Atlantic route, linking the major industrial regions on the east coast of North America with Western Europe, and within these regions lie half the world's major ports. At one time there was much passenger traffic along this route, but this has now greatly declined owing to competition from the airlines. The Cape route and the Suez route link Europe with the Far East and Australasia, although the latter was impassable from 1967 to 1975 during the closure of the Suez Canal. Many oil tankers are forced to use the longer Cape route from the Persian Gulf to Europe since they are too big to pass through the Suez Canal. The Panama route links the eastern and western coasts of the USA, between which there is a considerable amount of traffic, and this route has also replaced the Cape Horn route for trade between Europe and the west coast of South America. Other routes are the Trans-Pacific (the longest of all, linking America with Asia) and the South Atlantic (Europe to Brazil and Argentina).

There are many kinds of ships in use today, ranging from small **coasters** and cargo boats to massive **bulk carriers** of over

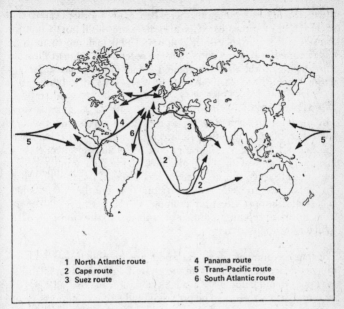

1 North Atlantic route	4 Panama route
2 Cape route	5 Trans–Pacific route
3 Suez route	6 South Atlantic route

Figure 49. World shipping routes

300,000 tonnes. General cargo is carried in **liners**, ships following fixed routes at regular times, and **tramp–steamers**, independent carriers which have no fixed schedule but sail when and where there is a cargo available. Many vessels are designed for specific cargoes, and the most important of these are those which carry bulk goods such as oil, grain and iron ore. Such ships are frequently of a very great size, since bulk carriage reduces transport costs. This principle can also be applied to much general cargo by grouping it in standard-sized containers, which can then be loaded into large, specialised **container-ships**. These large bulk carriers and container-ships require deep-water access and specialised handling equipment at the terminal ports, and therefore new berths have been built at some ports especially to deal with such vessels. For example, Milford Haven has facilities for oil traffic, Redcar for iron ore and Tilbury for grain and containers.

The largest merchant fleet in the world is that of Liberia, a small African state in which many foreign ships are registered owing to the less stringent regulations there. Panama is another example of such a 'flag of convenience', with a fleet out of all porportion to the size and importance of the country. The British merchant fleet has declined in relative importance in recent years, but still ranks third in the world after Liberia and Japan. Other major shipping nations include the USA, Greece, Norway, Holland and France.

Air transport

The most recent form of transport is by air, a development which has taken place in the twentieth century and in particular since the Second World War. The rate of technological advance has been very rapid, with the application of the jet engine and the building of larger and faster passenger aircraft. It is notable that this often has the effect of restricting the number of possible routes and airport sites: for example, the new Concorde supersonic airliner presents a number of problems to the countries over which it is proposed to fly.

The big advantages of air transport are its high speed and its independence of conditions on the ground. Provided that facilities for take-off and landing are available, aircraft can cross all types of land and sea with ease. The provision and siting of airports is, in fact, one of the main problems of air transport, since it is difficult to find suitably large areas of flat ground near enough to the main urban areas, which provide the traffic. Although automatic devices are beginning to solve the problem it is also desirable that airport sites should be in areas not subject to adverse weather conditions. Dense fog sometimes forces the closure of Heathrow Airport for many hours on end, and the delays which are caused by the diversion of aircraft to other airports negate the speed advantage of travelling by air.

Since the costs of movement by air are very high in comparison with sea transport it is quite uneconomic to carry bulky and low-value cargo by air. The most important groups of traffic are passengers, mail and small-bulk, high-value goods where speed of delivery is an important factor. However, in some parts of the world, remote areas are dependent upon aircraft for all kinds of supplies, especially in interior Africa and Australia and Arctic Canada where other forms of transport are impracticable. The most dense networks of commercial air services are in North America, Europe and the Mediterranean, and on the transatlantic route where passenger traffic has been captured from the shipping lines.

Heathrow Airport in London is the world's busiest international airport, other important international route centres being New York, Cairo, Karachi and Bangkok. The two main categories of passengers using air transport are government and business personnel and an increasing number of tourists, the latter frequently using **chartered aircraft** rather than **scheduled flights**.

Key terms

Coaster A small cargo vessel used for carrying goods along inshore routes.

Container-ship A modern cargo vessel especially designed for the transport of goods in containers. The use of such containers greatly reduces the cost and time of loading and unloading.

Freightliner A special fast train designed for the movement of goods, often in containers, over long-distance or inter-city routes.

Liner A large vessel used for the movement of passengers along fixed routes at regular times.

Motorway A relatively new arterial road designed for fast motor traffic, and with limited access from pre-existing roads.

Tramp-steamer An independent cargo vessel with no fixed schedule or route, sailing when and where cargoes are available.

Index

186

Examination Hints

Preparation

It is important that students are familiar with the syllabus for the examination for which they are entered. The syllabus usually provides information about the extent of the work to be covered, the amount of detail required, the approach to be adopted, and in some instances specifies certain textbooks as being particularly useful for the examination preparation.

Once the extent of the syllabus has been ascertained, the work involved should be divided into sections and a programme of preparation decided upon. This programming of work is most important. Students often underestimate the work involved in preparing for a GCE O-level or CSE examination, and many find themselves with an excessive amount of new material to be covered in the immediate pre-examination weeks. Under these circumstances certain parts of the work receive inadequate attention. While it is possible to learn the factual content of many courses in a relatively short time, a long period of steady work is essential in order to gain the deeper understanding and appreciation of the work which is necessary for high examination grades.

Consideration should be given to the aims of the examination. The study of geography involves more than the learning of factual information, and most examinations require more than the recital of elementary facts about products or regions. Such material should be a starting point, and continued study should lead to an appreciation of the relationships between the various branches of the subject. Personal observation and recording of data is the best method of study not only for fieldwork but for many branches of geography. Although this method is not a practical possibility for large parts of the work, it is nevertheless true to say that a certain amount of fieldwork is valuable whether this is specifically required by your syllabus or not. Some examining boards require the submission of 'field projects' or 'local studies' as an integral part of the examination. If so, sufficient time and effort should be devoted to the preparation of this work, for it often carries a substantial percentage of the total examination marks. Holiday visits either in this country or overseas, studies of your local district, excursions to local farms or factories, all of these can add a valuable depth of understanding and realism to your work which cannot be obtained from textbooks. As a second-best to fieldwork, frequent reference should be made to large-scale topographical maps covering contrasting types of landforms and settlement patterns.

Information must always be kept as up-to-date as possible. The content of geography as a subject is constantly changing as new factories and power stations are opened, harbours extended, new towns built or trade agreements drawn up. It is therefore important to read widely and to make a note of any relevant developments described in newspapers or magazines. Your statistical information should be as recent as possible. A useful reference book is *Geographical Digest*, an annual publication which gives up-to-date population and production statistics.

Notes made in class or extracted from textbooks should always be written up with care. Your notes must be neat, legible and concise, with clear, tidy sketch maps and diagrams, and a system of numbered points and headings to clarify and emphasise the main sections and divisions of the work. Remember that notes made in the early period of preparation must be completely clear when referred to many months later.

Attention must also be paid to the techniques of essay writing. Your knowledge, however good or bad it may be, will ultimately have to be expressed in the form of essay answers. The skills of essay-writing are therefore vitally important to any examinee. Frequent practice is essential. The selection and organisation of information and its clear and logical presentation in essay form must be thoroughly mastered. During the later stages of preparation you should practise essay-writing under the same pressure of time that the examination will impose. Access to previous examination papers is also useful at this stage so that you become familiar with the style and content of questions set in recent years. It is useful to be aware of the topics which are repeatedly examined, but unwise to plan your revision programme solely on the assumption that these topics will again be repeated. Attempts to restrict revision to a series of 'spot' questions can be very dangerous.

Revision should be a continuous process. It is no use leaving September's work until the following April before looking at it again. Basic facts should be constantly subjected to self-testing.

During the final period of intensive revision different methods may be used. Some students prepare a series of sketch maps on to which they compact a great deal of basic information. Others produce notes like those in the Key Facts Revision Section at the end of this book. The system adopted is a matter of personal preference. The important point is that you have a system of revision, and also that you remember that factual information alone will not satisfy the examiner.

The examination

It is probably true to say that as many candidates fail examinations through lack of examination technique as through lack of factual knowledge. No matter how much preparation, revision and learning of facts is undertaken, it is the quality of the answers composed in the examination room upon which success or failure depends.

The correct equipment is vital in a subject in which sketch maps, diagrams and illustrations are often of the greatest significance. Apart from pens and a ruler, the use of a variety of coloured pencils or coloured fibre-tipped pens will be of value for sketch maps and diagrams. For map-reading questions a set-square and protractor may be required, and a length of cotton or an accurate opisometer (if allowed by the examining board) should be employed for measuring distances on maps.

The well-prepared candidate will be familiar with the type of questions set in previous years. However, it is very important to read the instructions or rubric at the head of the question paper since minor changes do occur from time to time. Check how many questions are to be answered and calculate how much time should be allowed for each question. On some papers the map-reading question may carry extra marks. If so, it may be advisable to allow a little extra time for this. Bear in mind that time is needed at the beginning of the examination for the selection of questions, and again at the end for reading through your script and correcting minor errors. Above all, answer the correct number of questions.

The selection of questions is an important step. Think carefully before making a decision. Frequently a question which is difficult at first sight turns out to be far easier after a little consideration. Many candidates go straight for the 'easy' questions and often the standard required on these is higher than that needed on the more 'difficult' questions.

Where questions involve the composition of an essay presenting a reasoned description or argument, it is absolutely essential to plan carefully before writing the answer. The plan may be very brief, a series of paragraph headings together with a few of the main facts or points, but its real importance lies in the opportunity it gives for the arrangement of material in an orderly fashion. A logically presented essay with a minimum of fact will score more marks than a jumble of disconnected information.

Perhaps the greatest difficulty besetting the weaker candidate is the problem of relevance. This can only be overcome by planning carefully and by constantly keeping in mind the wording of the question. A simple, direct, concise style of presentation is required. No examiner is very happy with a script in which he has to search for each answer amongst a mass of irrelevant material, and such a script is never awarded very high marks.

It is important in geographical essays to avoid writing answers composed wholly of abstractions or generalisations. When presenting an argument it is essential to illustrate your work by the detailed quotation of relevant examples. On O-level papers this particularly applies to questions on physical topics and to some of the more general questions on human geography. Thus, if a question on the landforms of an area of glaciated highland is attempted, the description and explanation of a particular area such as part of the Lake District will score highly. If you can show familiarity with a topic through fieldwork, so much the better.

Diagrams and sketch maps should be incorporated in many O-level answers. Where a definite requirement for these is shown, a fixed percentage of the possible marks will be awarded for the illustration. Keep your maps simple, clear and relevant. Do not complicate them with detail which has no bearing on the question. The use of colour by pencils or fibre-tipped pens is of advantage if used with discretion, but the use of too much colour can conceal relationships it was intended to reveal. Keep to a scheme for the use of colour. Thus, blue or black might be used for rivers and coastlines, brown shading for upland, red for main roads and so on. Names should be printed on maps, never written. Sketch maps should have a margin or frame, and always carry a title and key. Although neatness and clarity are important, speed is also essential. If too much time is taken, then the written part of the answer is bound to suffer. Use the sketch map as a basis for your answer and refer to the points revealed by it, but do not waste time describing what is self-evident on the map. For diagrams, neatness, accuracy, clarity and relevance are again the cardinal virtues. Most diagrams are improved, and further knowledge revealed to the examiner, by careful annotation.

Finally, mention should be made of the importance of writing legibly and in good English. The examiner's impression of a candidate may often be strongly influenced by the relative ease with which the script may be read. A good standard of English (style, grammar and spelling) may likewise improve your chances.

The candidate's whole aim is to satisfy the examiner . . . upon this depends success or failure.

Suggested further reading

For certain examinations it may be necessary for students to have more detailed information on particular topics than is provided by this book. Also, as mentioned earlier, no attempt has been made here to include material on regional geography. The following booklist is included as a guide to additional sources of information.

Mapwork and field studies

F. C. Evans: *Map Study*, Oliver & Boyd, 1972.
P. A. Sauvain: *Facts, Maps and Places*, Hulton, 1970.
F. R. Dobson & H. E. Virgo: *A New School Geography* (Book III, 'Map Reading and Local Studies'), English Universities Press Ltd., 1964.
J. Haddon: *Local Geography: Geographical Surveys in Rural Areas*, Philips, 1964.

Physical geography

R. B. Bunnett: *Physical Geography in Diagrams*, Longman, 1965.
D. C. Money: *Basic Geography*, University Tutorial Press, 2nd edition, 1972.
P. Speak & A. H. C. Carter: *Sketch Map Geography* (Book I, 'Physical Geography'), Longman, 1967.

Human geography

J. Dawson & D. Thomas: *Man and His World: An Introduction to Human Geography*, Nelson, 1975.
N. J. G. Pounds: *Introduction to Economic Geography*, Murray, 1951.

O. Hull: *A Geography of Production*, Macmillan, 1968.

M. Storm: *Urban Growth in Britain*, Oxford University Press, 2nd edition, 1975.

Regional geography of the British Isles

D. M. Preece & H. R. B. Wood: *The British Isles*, University Tutorial Press, 18th edition, 1974.

N. J. Graves & J. T. White: *Geography of the British Isles*, Heinemann, 1971.

J. H. Lowry: *The British Isles*, Arnold, 3rd edition, 1968.

P. Speak & A. H. C. Carter: *Sketch Map Geographies* (Book II, 'The British Isles'), Longman, 1967.

A. R. Tolson & M. E. Johnstone: *A Geography of Britain*, Oxford University Press, 2nd edition, 1972.

K. B. Stevenson: *Geography of the British Isles*, Blandford, 1973.

Regional geography of Europe

J. H. Lowry: *Europe and the Soviet Union*, Arnold, 2nd edition, 1972.

F. J. Monkhouse: *Europe* in *Geographies: A Certificate Series*, Longman, 4th edition, 1975.

D. M Preece & H. R. B. Wood: *Europe*, University Tutorial Press, 12th edition, 1969.

Regional geography of North America

A. W. Coysh & M. E. Tomlinson: *North America*, University Tutorial Press, 16th edition, 1975.

E. W. Young: *North America*, Arnold, 2nd edition, 1970.

In addition to these texts a good atlas is essential for any geographical study. To a large extent the choice of atlas is a matter of personal preference. However, the following atlases are particularly useful for the O-level student:

Atlas Three, Collins/Longman, 1973.

The Oxford Home Atlas, Oxford University Press, 3rd edition, 1960.

Newnes International World Atlas, Newnes, 1967.

Key Facts

Revision Section

Map reading
Conventional signs
Learn the symbols used on the 1 inch: 1 mile (7th series) and the 2½ inch: 1 mile OS maps.

Techniques of map reading
(1) **Grid references** Four-figure references should be used for large features and six-figure references for small features. Remember that eastings (top margin) are given before northings (side margin). (2) **Measurement of distances** For winding routes such as roads, rivers, railways, etc., use either a pair of dividers to 'step off' sections of the route, or the straight edge of a piece of paper, or follow the route with a length of cotton. Convert the map distance to ground distance by means of the scale line on the map. (3) **Calculation of average gradient** Find the vertical difference in height between the two points and also the horizontal distance between them. Divide the horizontal distance by the vertical difference and express the result as 1 in x. Always show full working. (4) **Measurement of bearing** Measure the angle between grid north and a line to the feature concerned in a clockwise direction. Always express the result as a full-circle angular bearing only, e.g. 326°. (5) **Section drawing** Hold the straight edge of a piece of paper along the line of the section and mark off all contour intersections. Transfer these points to a sheet of graph paper at the appropriate height. Use a vertical scale of $\frac{1}{10}$ inch to represent 50 feet for the 1-inch map and $\frac{1}{10}$ inch to represent 25 feet for the 2½-inch map. Keep the horizontal scale the same as for the original map. Join up the points with a smooth curve as far as possible. Annotate the section with neat lettering and vertical arrows to the appropriate features. Add a statement of vertical and horizontal scales. **Sketch sections** Follow a similar method to the above but plot only those contours at the top or bottom of even slopes. Sketch sections are required in questions dealing with intervisibility. **Determination of intervisibility** involves drawing a straight line on the section between the object and observation point. If the line is clearly above the section line at all places then the two points are intervisible. (6) **Measurement of areas** For the measurement of areas from the 1 inch: 1 mile map transfer the outline of the feature on to $\frac{1}{10}$ inch square graph paper. Count the number of complete squares and estimate the aggregate of part-squares contained within the outline. Divide the total by 100 to convert to square miles. Great care and accuracy are essential in all questions involving map measurement, section drawing, etc.

Map reading: description

Answers to questions involving the description of features of the physical or human landscape must be presented in a logical ordered manner. Avoid writing a series of haphazard and disconnected observations. Remember to use map evidence only, and to give grid references to specific features mentioned in any description.

Description of uplands Mention location, size, average height and maximum height. Consider the pattern and degree of dissection by stream valleys. Examine the steepness of slopes. Using map evidence of permeability, quarries, place-names, etc., suggest the rock type if possible. Draw attention to special features such as those caused by glaciation.

Valley descriptions should mention the steepness of the valley sides, width of the valley, the downstream gradient and valley cross-profile. An assessment should be made of whether the valley is in a youthful, mature or old-age stage, or whether it has been modified by rejuvenation or glaciation.

Description of coasts should include reference to the alignment of the coast. Examine the degree and nature of indentation (bays, coves, headlands, etc.). Describe erosional and depositional features (where appropriate), and comment on the size of features mentioned in the description, e.g. the height of cliffs and the width of beaches.

View from a point In descriptions of the view from a specific point, assume a 30° angle of vision (15° on either side of the line marking the direction of vision). Describe features either from left to right or from foreground to distance.

Complete physical analyses Sub-divide the map sheet or map extract into contrasting physical areas. Describe each in turn according to the methods outlined above. A sketch map is essential for this type of question. It should show the approximate boundaries of each area and a descriptive title should be provided for each.

Map reading: interpretation

Map interpretation is concerned with the explanation, analysis, and relationships between patterns of relief, vegetation, settlement, communications, etc. shown on a map extract. Reference should be made to map evidence only. In many cases map evidence will be suggestive rather than conclusive. Nevertheless, additional information or knowledge about the map area should not be introduced. Organise the analysis in a clear logical manner. Analysis of the distribution of **vegetation patterns** should include reference to the requirements of the vegetation concerned, and also to the reasons why the land is not used for alternative purposes. Thus, an analysis of the distribution of woodland in an area should make reference to relief, gradients, rock-type, soils, water-supply, drainage, etc. These may help to explain why various areas have been given over to woodland and not used for arable or pastoral farming.

Analysis of **settlement patterns** is concerned fundamentally with an assessment of the possibilities of making a living and establishing settlement in an area. This will involve reference to relief, drainage, water-supply, soils, mineral resources, communications, etc. The site and position of farms, hamlets, villages and towns have been chosen for various reasons, some of which may be evident from the map, e.g. village sites may have been determined by shelter, availability of water-supply, safety from flood risk, accessibility to different types of farmland, etc. while market towns have often developed because of accessibility from a wide surrounding area.

Site is the land on which a settlement is built, e.g. a river terrace, while **position** is the relationship of the settlement to its surrounding area. The latter involves consideration of natural route-ways.

Analysis of **communication patterns** usually involves consideration of the relationships between relief and roads and/or railways. A sketch map is essential for showing the arrangement of upland areas, gaps, valleys, bridging points, etc. and the main lines of communications. Railways can only negotiate gentle slopes (maximum c. 1 in 80). In this connection note the use of embankments, cuttings and tunnels. Roads are far less affected by slopes, although gradients exceeding 1 in 7 are rare on main roads. Canals are most restricted by factors of relief.

Map and photograph

Questions are frequently set which involve correlation of an aerial photograph with an OS map (usually an oblique photograph). Questions usually relate to one of the following: (1) measurement of the direction and angle of view of the photograph; (2) orientation of the map-sheet and identification of features on the photograph; (3) determination of the time of day when the photograph was taken.

Field studies

The importance placed on local studies varies from one examining board to another. Be sure you know your syllabus requirements. Whether your syllabus specifies such work or not, it is important to have studied a small area in detail in the field. This type of work provides the best method of studying the relationships between various geographical factors. Such specific information can be profitably introduced into answers to many general questions. The type of material that might be included in a field study is listed below.

(1) **Location and delimitation of the study area**

(2) **Geology, relief and drainage** Consult the geological map. Study rock outcrops and collect rock specimens. Observe and describe the relief and drainage features of the study area. Map and sketch river features.

(3) **Weather records** Collect data on temperature, rainfall, pressure and winds.

(4) **Soils and vegetation** Examine soils in the study area. Draw maps to show woodland, common-land, etc. Note species of plants and trees. Attempt to relate their distribution to relief, soil, drainage, etc.

(5) **Land use** Examine soils. Map the distribution of crops on $2\frac{1}{2}$ inch or 6 inch: 1 mile base maps. Relate land use to climate, relief, soils, markets and other relevant factors. Sample studies of individual farms provide useful information to illustrate crop rotations, the annual work pattern, methods of marketing etc. (See page 22, on sample farm studies.)

(6) **Industry** Map the distribution of industrial sites in the study area. Attempt a classification of the industrial premises revealed. Relate different types of industry to raw materials, power, labour-supply, communications, markets, etc. (See page 23, on sample factory studies.)

(7) **Settlement** For rural areas, map the distribution of different types of settlements. Examine the services provided in each, and the extent of the areas served. Urban studies should include the mapping of different urban functions (housing, industry, retailing, etc.), analysis of street patterns, building types, building materials, etc. Historical information should be introduced only if it helps to explain the growth of the settlement on elements of the present pattern.

(8) **Communications** Produce a series of maps to show roads, railways, canals, etc. Relate the patterns to natural features and settlements. Traffic flows may also be usefully mapped.

Emphasise those themes which are 'special' to your study area, and determine its own particular geographical character.

The planet earth

Latitude and longitude

Latitude is the angular distance of a place north or south of the equator from the centre of the earth. The latitude of a place is determined by subtracting the angle of the noon sun above the horizon on the equinox from 90°. Lines of latitude (parallels) run east to west and are parallel to the equator. Lines of latitude are a constant distance apart (about 110 km).

Longitude is the angular distance of a place east or west of the **Greenwich meridian**. The angle is measured at the axis of the earth. Lines of longitude (meridians) run from the North to the South Pole.

Longitude and time

The earth rotates from west to east through 360° in 24 hours, i.e. 15° in 1 hour, or 1° in 4 minutes. Thus, local time will be later to the east of a place and earlier to the west. Be able to calculate the local time at one place given local time at another, e.g. The time in New York (74°W) is 12 noon: what is the local time in Moscow (38°E)?

Difference in longitude	=	112°
Difference in time	=	112/15 hours
	=	7 hours 28 minutes
Therefore, time in Moscow	=	7·28 p.m.

Time zones To eliminate small time differences over short distances the world is divided into a number of time zones (usually 15° in extent). Watches must be adjusted by one hour when passing from one time zone to another.

The International Date Line follows the 180° meridian, with slight deviations round island groups, etc. When crossed from east to west a day is lost; when crossed in the reverse direction a day is repeated.

Great circles

The shortest distance between any two points on the globe lies on a line, the plane of which cuts through the centre of the earth. Such lines are sections of what are termed great circles. For east/west routes in the northern hemisphere a great circle forms an arc north of the line of latitude connecting the two places; in the southern hemisphere the great circle route forms an arc to the south of the appropriate line of latitude.

The seasons

The earth takes 365¼ days to complete one revolution round the sun. For convenience a year is taken as 365 days with a leap year of 366 days every fourth year. The earth's axis is inclined at an angle of $23\frac{1}{2}°$ to the perpendicular or $66\frac{1}{2}°$ to its plane of orbit (path around the sun). The overhead sun therefore appears to move from the **Tropic of Cancer** ($23\frac{1}{2}°$N) on 21 June to the **Tropic of Capricorn** ($23\frac{1}{2}°$S) on 22 December, i.e. the two tropics mark the northern and southern limits of the overhead sun.

The higher the sun in the sky the greater the amount of heat energy received. Mid-summer in the northern hemisphere occurs in June, while in the southern hemisphere this is the period of mid-winter. Spring and autumn seasons relate to the intermediate periods between these two extreme positions of the overhead sun.

Length of daylight each day also varies according to the position of the overhead sun. The northern hemisphere has long nights and short days in December. The reverse occurs in the southern hemisphere. A useful exercise is to compare sunrise and sunset times for places in northern Scotland and southern England in summer and winter. North of the **Arctic Circle** 24 hours of daylight are experienced on 21 June (**the midnight sun**), while at the same time areas south of the **Antarctic Circle** experience continuous darkness. On 22 December conditions are reversed.

Solstices
21 June (the summer solstice in the northern hemisphere): on this date the sun is overhead at the Tropic of Cancer.
22 December (the winter solstice in the northern hemisphere): on this date the sun is overhead at the Tropic of Capricorn.

Equinoxes
21 March and 23 September The sun is overhead at the equator on these two dates. All places therefore experience 12 hours' daylight and 12 hours' night on these two dates with sunrise and sunset at 6.00 a.m. and 6.00 p.m. local sun time.

Be sure that you know and are able to draw diagrams to illustrate the position of the earth and the sun and the length of daylight on these four dates.

Rock types
The rocks of the earth's crust may be classified in various ways, e.g. according to their mineral composition, age, mode of formation, etc. On the basis of mode of formation, three main groups may be identified, namely, igneous, sedimentary and metamorphic rocks.

Igneous rocks

These are the result of the cooling of molten material or **magma** from deep below the earth's surface. If the magma cooled at depth the rocks are termed **plutonic** or **intrusive**. These are generally coarse-grained rocks with large crystals, e.g. granite. Rocks resulting from the cooling of magma on the surface are termed **volcanic** or **extrusive**. These are fine-grained rocks, e.g. basalt.

Intrusive igneous rocks which have been injected into the surrounding rocks assume many different forms. These include **bathyliths**, **dykes** and **sills**. Generally these are only evident after the overlying rocks have been removed by erosion.

Sedimentary rocks

Most sedimentary rocks are derived from the erosion and weathering of existing rocks and the movement and deposition of the resulting sediment. As layer upon layer of sediment is deposited it becomes consolidated and cemented by pressure and chemical action, e.g. the formation of clay, shale and sandstone. These are termed **clastic** sedimentary rocks. Other sedimentary rocks are formed by **chemical** action, e.g. rock-salt and gypsum. Others are of **organic** origin, e.g. carboniferous limestone and coal. Sedimentary rocks are typically stratified. **Strata** are separated by **bedding planes**, while **joints** tend to run perpendicular to the bedding.

Metamorphic rocks

These are former sedimentary or igneous rocks which have been transformed by great heat and/or pressure. This may have resulted from involvement in earth movements or contact with magma. The form of metamorphic rocks is determined by a number of factors, including the nature of the original rock, the type of metamorphism to which it has been subjected, and the intensity and duration of the metamorphism. Examples of metamorphic rocks include quartzite, slate, schist, marble, gneiss, etc. In many cases minerals have been flattened and rearranged in roughly parallel bands which run through the rock. This is known as **foliation**.

The **properties** of the most common rocks should be known, together with their **distribution** in the British Isles. The distinctive **scenery** associated with these rocks should have been studied, either in the field or from OS maps and photographs. Know **sample areas** of the main rock types and be able to draw maps and diagrams to illustrate structure, relief, drainage, etc. for areas of granite, carboniferous limestone, chalk, etc.

Granite

Granite is a resistant, crystalline rock composed of mica, quartz and felspar. It usually forms mountain or moorland areas, e.g. Dartmoor, Bodmin Moor, Cheviot Hills, much of the Grampian Highlands, etc. It is an impervious rock and therefore there are many surface streams. Soils are poor, thin and acid. Accumulations of peat may be found in poorly drained areas. Typical vegetation consists of rough grassland, peat bogs and scattered woodland. On Dartmoor projections of bare rock on hill summits are known as **tors**. Mineral workings are found in many granite areas, e.g. Cornish tin-mines.

Carboniferous limestone

The typically dry, upland landscape of Carboniferous limestone areas is known as **karst scenery**. This results from the permeability and the solubility of the limestone. Notable examples of limestone scenery are found in the Mendip Hills, the Peak District of Derbyshire, the Malham district of Yorkshire, etc. In these areas the surface may be furrowed by solution to form **clints and grykes** (e.g. on the slopes of Ingleborough, Yorkshire). **Swallow-holes** may be found, some carrying streams underground to **cave systems** in which **stalactites** and **stalagmites** may occur. These are caused by the re-deposition of calcium carbonate from percolating water. **Resurgent streams** (e.g. the River Aire at the foot of Malham Cove) and **springs** are commonly found at the base of the limestone outcrop. **Dry valleys** (often gorge-like) and exposures of bare rock known as **scars** are common features. Soils are generally thin and alkaline. Typical vegetation consists of rough grassland with many bare rock exposures.

Chalk

This is a soft type of limestone with less well-developed jointing. Chalk areas are usually characterised by a **smooth, rounded type of relief**. Areas include the North and South Downs, the Chiltern Hills, the Lincolnshire and Yorkshire Wolds, etc. **Dry valleys** and **winterbournes** are common features of the landscape. Land use is very varied. Traditionally chalk areas are given over to sheep grazing, but in recent years there have been notable extensions of arable cultivation into chalk upland areas. Soils are usually thin and dry. **Dew ponds** have been constructed in many areas to overcome the lack of surface water. Patches of **clay-with-flints** give deeper soils, especially on hill summits, and often support a cover of beech woodland.

Millstone grit

Millstone grit is a dark coloured, coarse, resistant sandstone. Millstone grit outcrops usually form areas of high moorland. Many of the hills have sharp ridges and steep sides often known as '**edges**'. The typical vegetation in such areas consists of coarse grass, bracken, heather and cotton grass. **Surface streams** are abundant and in places **peat moors** and bogs are found. **Reservoirs** giving soft water, have been built in many millstone grit areas. Typical gritstone scenery may be seen in the Central Pennines west of Sheffield.

Earth movements

Earth movements may be divided into two main groups: namely, epeirogenic and orogenic movements. **Epeirogenic movements** are relatively slow movements involving the broad uplift or submergence of extensive areas. The rocks involved are typically tilted or warped rather than intensively folded or fractured. **Orogenic movements** are more intensive and generally produce complex folding and fracturing of the rocks involved. These are the forces responsible for the creation of the world's major mountain belts.

Earthquakes

These originate in areas of crustal instability (similar to areas of volcanic activity). The **epicentre** is the point on the surface immediately above the **focus** of the earthquake. Damage results from **seismic waves** which spread out from the epicentre. Sea waves generated by earthquakes are known as **tsunami**.

Fold structures

Pressures in the earth's crust acting along a horizontal axis may produce folds in sedimentary rocks. Upfolds are termed **anticlines**, downfolds **synclines**. Variations in the form and intensity of fold structures may be noted. These include **simple folding, asymmetric folding, recumbent folding**, and **nappe structures**. The latter also involve low-angle thrusting. **Geosynclines** are enormous downwarped troughs in which sediment accumulates prior to being thrust up into fold structures.

Fault structures

In some cases rocks respond to pressure by fracturing and moving along such a fracture line. This is termed **faulting**. Different types of fault may be noted, e.g. **normal fault**, and **reverse fault**. Both will produce a **fault-scarp** at the surface. More complicated fault structures involving numerous faults include **rift valleys, horsts**, and **block mountains**.

Volcanic activity is often associated with both folding and faulting.

Volcanoes

Two main types of volcanic eruption may be noted: namely, **crater eruptions** and **fissure eruptions**. In the case of a fissure eruption the lava is extruded through a long fissure to form a lava plateau, while in the case of a crater eruption a volcanic cone results. Volcanic cones vary according to their age, the violence of earlier eruptions, and the type of lava being emitted. Fluid **basic lava** forms a **shield volcano**; viscous **acid lava** forms a compact steep-sided cone; a **composite cone** is constructed of alternate layers of lava and volcanic ash and cinders; a cone partially destroyed by a violent eruption is known as a **caldera**.

Volcanoes may also be classified as **active, dormant** and **extinct**. The world distribution of volcanoes should be learnt. Zones of volcanic activity include the 'Fiery Ring of the Pacific', the Mediterranean basin, East Africa, the West Indies, Hawaii, etc. **Hot springs** and **geysers** are found in many of these regions.

Weathering and soils

The term **weathering** refers to the decomposition and **disintegration of rocks** *in situ*. Do not confuse the terms weathering and erosion. **Physical weathering** takes place when a rock is reduced to smaller fragments without undergoing any chemical change. Such processes include **frost-shattering** and the break-up of rocks through alternate heating and cooling (**exfoliation** and **granular disintegration**). **Chemical weathering** produces rock materials which are chemically different from the parent rock. The processes of chemical weathering include **oxidation, hydration, carbonation** and **solution. Biological weathering** refers to the break-up of rocks through the action of plant roots, burrowing animals, etc.

Soils

Climate, vegetation and geology all influence soil development. Soils are related to parent material in that the weathering of rocks results in the formation of loose surface material in which plants may grow and soils develop. Climate influences soils through the vegetation it controls, and through the moisture that is available for processes such as leaching, etc.

Soil composition Soil contains weathered parent rock, humus or decaying vegetation, air, water, plant roots and fauna such as earthworms. All are important in soil formation.

Description of soils Soils may be described and classified in various ways. (1) By their **texture**, i.e. by the proportion of clay, silt and sand particles they contain. (2) By their **acidity**: pH values indicate the amount of hydrogen in the soil solution (neutral soils have a pH value of 7, values above and below 7 indicate basic and acidic soils respectively). (3) By their **profile**: a soil profile is simply a vertical section through the soil from the ground surface to the parent rock; three basic horizons may be identified in most soil profiles.

The zonal concept of soil development This is based on the idea that regardless of the parent material (with a few exceptions, such as limestone) soil types will develop according to the climatic conditions in which they exist. The following zonal soils should be known together with their typical profiles: (1) **podsol soils**; (2) **brown forest soils**; (3) **chernozem soils**; (4) **lateritic soils**. Intrazonal and azonal soils do not fit into the system of zonal soils. **Intrazonal soils** develop where special factors (e.g. poor drainage) prevent normal development. **Azonal soils** are young soils which have had insufficient time to develop, e.g. soils on recent alluvial deposits.

Rivers, valleys and lakes

Fluvial processes

The seasonal variation in the volume of water in a river is called its **régime**. As rivers flow they perform the interrelated functions of transport, erosion and deposition.

Transport Rivers move material in three ways: (1) **in solution**; (2) **in suspension**; (3) **by rolling fragments** along the river bed. The transporting power of any river depends on the volume and speed of flow, the nature of the stream channel, and the type of material being carried. Transport is most effective in times of flood.

Erosion A river erodes its bed by four processes: (1) **hydraulic action**; (2) **corrasion**; (3) **attrition**; (4) **solution**.

Deposition For any given volume and speed of river flow there is a maximum load that can be carried. If the current is checked, material will be deposited. The largest fragments are dropped first, the finer material later, i.e. a natural grading of the deposited material takes place.

The development of a river valley

Rivers and their valleys often display three stages of development: namely, youthful stage, mature stage and old-age stage. However, many variations from this simple threefold division are possible.

Youthful stage The source of a river may be a spring, lake or glacier. At this stage the volume is small, the gradient steep, and the flow rapid. Thus, downcutting is dominant, but rainwash, soil-creep and tributary streams all contribute to valley widening. The valley forms a steep-sided, V-shaped section at this stage. **Interlocking spurs** may be formed.

Mature stage The volume of water is larger at this stage, but the gradient is less steep and therefore the flow is slower. **Valley widening** becomes more important. As **meanders** form and migrate downstream they produce the beginning of a **flood-plain** with **river-cliffs**.

Old-age stage Volume is at a maximum, but gradients are very slight, and therefore flow is very slow. A **braided stream channel** may result from the imperceptible gradient. **Deposition** is more important than erosion at this stage. The river channel (sometimes with **levées**) may be raised above the flood-plain level by the deposition of **alluvium**. Complex meander forms with **ox-bow lakes** are common. Wide flat flood-plains with bordering **river bluffs** are also typical.

Deltas

Two main types may be noted: (1) **arcuate deltas**, e.g. the Nile delta; (2) **Bird's-foot deltas**, e.g. the Mississippi delta. Many fail to fall into either category, e.g. the Rhine delta. Deltas form extensive flat areas close to sea-level with the main river splitting into numerous **distributaries**. The following conditions are involved in delta formation: (1) large supply of silt; (2) river current checked by the sea (salt-water coagulates fine silt); (3) gently shelving offshore gradient; (4) weak tidal range. Note, however, that deltas **can** form on coasts with a large tidal range provided that the river deposits more silt than is swept away, e.g. the Ganges delta coast has a tidal range of 5 metres.

Waterfalls

Various causes may be noted: (1) bands of hard rock across the river's course: the fall retreats slowly upstream by the undercutting of the resistant bed, thus forming a gorge, e.g. Niagara Falls; (2) hanging valleys in glaciated regions; (3) rock-steps in glaciated valleys, etc.

Drainage patterns

Three main types of drainage pattern may be noted. However, the variety of stream patterns is enormous, and many fail to fall into any one group. (1) **Radial drainage**: streams radiate from the centre of an upland area, e.g. Dartmoor. (2) **Dendritic drainage**: this is associated with areas of uniform rock with no marked controls on river development; rivers and streams form a branching, tree-like pattern. (3) **Trellis drain-**

age: this tends to develop in areas of gently dipping sedimentary rocks of varying resistance. Main rivers (consequents) flow across the various outcrops cutting gaps through the ridges. Subsequent tributaries excavate the weaker outcrops. The river pattern therefore assumes a rectilinear form. **River capture** may take place to produce a sharp bend in the course of the river (**elbow of capture**) with a **wind gap** and **misfit stream**.

Lakes

Lakes vary very greatly in size from a few hectares to thousands of square kilometres. A broad distinction may be made between fresh-water lakes with an outlet stream, and saline or salt-water lakes with no outlet. Most lakes belong to the former category. Various types of lakes may be distinguished according to their mode of formation. These include:

(1) **lakes produced by erosion**, e.g. corrie lakes, ribbon lakes, lakes occupying hollows owing to solution, etc.;

(2) **lakes produced by deposition**, e.g. barrier lakes pounded up behind lava flows, moraines, deltaic deposits, etc.;

(3) **lakes caused by earth movements and vulcanicity**: this group includes lakes occupying hollows resulting from the tilting, folding or faulting of strata, as well as crater lakes.

Glaciation

At the present time glacial activity is confined to high-latitude and high-altitude areas, e.g. Greenland, the Alps, etc. During the Quaternary Ice Age glaciers and ice-sheets affected much larger areas than at present. The results of ice action may thus be seen far beyond the limits of present-day glaciation, e.g. Britain lay under the ice as far south as a line roughly corresponding with the River Thames. Three types of ice accumulation may be noted: namely, **valley-glaciers, piedmont glaciers** and **ice-sheets**.

Mountain glaciation

If temperatures are low enough more snow may collect on mountain sides than is lost by melting. Hollows in which the snow accumulates are deepened by freeze-thaw action; meltwater removes the rock fragments. As the snow deepens it becomes changed first into **névé**, and then into glacier ice which, under the force of gravity, moves out of the hollow and downhill to form a valley glacier. The original hollow, enlarged by the ice, is known as a **corrie**. Complex patterns of **crevasses** develop on the surface of glaciers owing to the different rates of movement of the various parts of the glacier and irregularities in the underlying bed. Several types of moraine are found along a typical valley glacier. These include lateral moraine, medial moraine, ground moraine, terminal moraine and recessional moraine.

Various landscape features resulting from glacial erosion in mountain areas may be noted. These include: (1) **corries**: deep hollows with steep back and side walls, often containing a small circular lake; (2) **arêtes**: formed by the reduction of the ground between two corries into a steep-sided narrow ridge; (3) **pyramid peaks**, which may result from the erosion of a mountain by a number of corries; (4) **U-shaped valleys**: a normal V-shaped valley may be modified into a U-shaped section by glacial erosion; (5) **truncated spurs and hanging valleys**, often seen along such glaciated valleys; (6) **rock-steps and ribbon lakes**, caused by the differential erosion of a valley floor by a former glacier; (7) **roches moutonnées**: ice-eroded rock structures, smoothed on one side by the ice and left rugged on the other side.

Lowland glaciation
In lowland areas glacial deposition is generally more important than erosion. Features of glacial deposition include: (1) **deposits of boulder clay or till**; such deposits may contain (2) **erratic boulders**; (3) **kettle-hole lakes**, frequently found on the surface of boulder-clay deposits; (4) **drumlins**, composed of boulder-clay moulded by the ice into clusters of low, elongated, rounded hills; (5) **outwash sands and gravels**, laid down by streams issuing from the front of a former ice-sheet or glacier; (6) **eskers**, winding ridges of sand and gravel laid down by sub-glacial streams; (7) **loess deposits**, found adjacent to many areas of former glaciation; they consist of wind-blown dust derived from the glacial deposits.

The effects of glaciation are not limited to the erosional and depositional features outlined above. In many areas affected by glaciation, rivers have been diverted into new courses and drainage patterns greatly modified. Examples include the modification of drainage in North Yorkshire and the diversion of the River Severn.

Deserts
Deserts and semi-desert areas cover about one-third of the earth's land area. Three types are recognised: namely, sand deserts (**erg**), stone deserts (**reg**) and rock deserts (**hammada**). Desert areas are characterised by a lack of cloud and therefore have a large diurnal range of temperature. Alternate heating and cooling of rocks causes them to break up as a result of either **granular disintegration** or **exfoliation**.

The action of the wind
The strong winds which are characteristic of deserts affect their landforms by erosion, transportation and deposition. Wind erosion is accomplished by the movement of sand grains which have an abrasive effect on rock structures and may lead to the formation of (1) **deflation hollows**; (2) **fretted and polished rock surfaces**; (3) **pedestal rocks**; (4) **zeugens**; (5) **yardangs** (gullies in soft rock).

Weathered material can be carried by the wind during sand-storms. Fine dust may be carried far beyond the deserts and deposited as **loess**. Note the different forms of dunes in sand deserts: (1) **attached dunes**; (2) **barchans** (crescent-shaped dunes); (3) **seif-dunes** (linear dunes).

The action of water
Although deserts have, by definition, under 250 mm of rain per year, water action is important. Occasional rain falls in violent storms and the ground is unprotected by vegetation. **Sheet and gully erosion** results, and **alluvial fans** form at the foot of the high ground. A number of alluvial fans may coalesce to produce a **bahada**. The **wadis** of the Sahara were probably formed in the past when the climate of the deserts was more humid than at present. In places a temporary shallow lake may form in a desert basin. This is termed a **playa, salina** or **shott**.

Resistent, isolated rock masses in deserts are known as **inselbergs** (**mesas** and **buttes** in the USA).

Coasts
Coastlines are undergoing constant changes owing to erosional and depositional processes caused chiefly by the action of **waves**. Waves originate in the open sea as a result of the wind. **Wave length** depends on the speed and duration of the wind, but **wave height** increases with the distance over which a wave has travelled (**fetch**).

Coastal erosion
Several processes are involved: these include the weight of water of breaking waves, the pressure of air forced into rock crevices, abrasion by rock fragments and shingle, and solution of certain rocks, e.g. chalk.

Features of coastal erosion include: (1) **cliffs**: their form depends upon the hardness and structure of the rocks; seaward dipping beds give less steep cliffs than beds dipping inland; (2) **wave-cut platforms**: with the recession of cliffs an eroded rock platform results; (3) **caves**: these are often related to jointing or faulting; (4) **narrow inlets (geos)** which may also be eroded along lines of weakness; continued erosion may produce (5) **blow-holes** or **gloups**; (6) **natural arches** (e.g. Durdle Door, Dorset), and eventually (7) **sea stacks** (e.g. the Old Man of Hoy, the Needles). Notice that where beds of varying resistance outcrop on a coast the harder rocks will form headlands and the softer rocks bays. This is referred to as **differential erosion**. The process is well illustrated by sections of the Dorset coast.

Coastal deposition

Beaches of sand and shingle will form where a supply of material is available. This may be on an exposed coast, but more usually in an embayment as a **bay-head beach**. Where a large area of drying sand is exposed at low tide and strong onshore winds prevail, the sand may be carried to the head of the beach to form **dunes**. Longshore drift (caused by oblique waves) is important in moving material along beaches. Where the coast changes direction or is broken by an estuary, a **sand-spit** may form, e.g. Orford Ness, Suffolk. If an embayment is sealed off by this process the feature is termed a **sand-bar**, e.g. Slapton Sands, Devon. A **tombolo** is a spit connecting an island to the mainland, e.g. Chesil Beach, Dorset. On very gently shelving shores the waves may build up an **offshore bar**. **Salt-marsh** may form in the sheltered waters on the landward side of these depositional features, e.g. the salt-marsh area on the landward side of Scolt Head Island on the north Norfolk coast.

Changes of sea-level

Sinking or uplift of the land or a rise or fall in sea-level also has an important effect on the nature of coastlines.

Coasts of submergence Various distinctive coastal features result from the submergence of a coastal area. (1) **Estuaries**, which may have been formed as a result of the submergence of a low-lying coast crossed by a major river. The lowest section of the valley has been flooded to form an estuary, e.g. the Thames estuary. (2) **Rias:** another form of drowned lowland coast. In this case, the sea has penetrated into the lowest sections of the valleys of a former dendritic drainage pattern to produce a series of branching arms of the sea often extending far inland, e.g. the Fal estuary, and Plymouth Sound. (3) **Fjords:** partially drowned glaciated valleys, hence their steep sides, great depth and distribution limited to glaciated regions. A typical feature is a shallow threshold near the mouth, e.g. Sogne Fjord. (4) **Dalmatian coasts**, which result from the submergence of the lower parts of fold-mountain ranges running parallel to the coast. The submergence produces a series of gulfs (often T-shaped) and long narrow offshore islands roughly parallel with the mainland, e.g. the coast of Dalmatia, Yugoslavia.

Coasts of emergence (1) **Coastal plains:** uplift of gently-shelving offshore areas may produce wide coastal plains, e.g. the coastal plain of the Atlantic coast of the USA. (2) **Raised beaches:** former wave-cut platforms terminating in an old cliff, but now lying above the present sea-level. Sometimes a cover of marine deposits may be found. These are local features, rarely continuous over long distances.

The oceans

The area of the oceans (70·8 per cent of the earth's surface) is far greater than the land area (29·2 per cent). The ocean floor may be sub-divided as follows: (1) **the continental shelf:** the gently-sloping sea-bed bordering the continents: it varies greatly in width but extends down to *c.* 100 fathoms, at which depth gradients increase and the shelf yields to (2) **the continental slope:** a relatively steep slope down from the continental shelf to the deep ocean floor; (3) **the deep-sea plain:** here depths of 2,000–3,000 fathoms are typical; submarine relief is varied; (4) **the deeps:** narrow, linear **troughs** of exceptional depth, which often lie close to island arcs and zones of earthquake and volcanic activity, e.g. the Mariana trench.

The waters of the oceans

Salinity   Variations in surface salinity are caused by the mixing and circulation of the ocean waters, evaporation rates, and the supply of fresh water into the sea. Highest salinity is found near the tropics owing to the high evaporation. It is lower at the equator owing to heavy rain and lower evaporation. Towards the poles melting ice supplies fresh water and lower values are typical.

Ocean currents   Ocean currents are caused chiefly by differences in the temperature, salinity and density of different water masses. Ocean '**drifts**' are caused by the wind and move in the direction of prevailing winds. The chief ocean currents and drifts should be known, together with their effects on the climate of the bordering lands.

Coral formations

Coral is formed from the lime secretions of the coral polyp which requires warm, clear, shallow seas. Corals are therefore restricted to seas between latitudes 30°N and 30°S. Three types of coral structure may be noted: (1) **fringing reefs**; (2) **barrier reefs**; (3) **atolls**.

Weather

The term 'weather' denotes the conditions of the atmosphere at a particular place and time. Weather refers to conditions at a specific point in time, and should not be confused with **climate**, which is the normal or average weather for a particular place.

Weather recording

Temperature This is measured by means of **maximum and minimum thermometers** kept in a **Stevenson's screen**. From maximum and minimum readings a daily mean can be calculated. From these daily means a mean monthly temperature is then calculated. Temperature maps use **isotherms** (lines joining places of equal temperature after correction to sea-level equivalents).

Precipitation The instrument used is a **rain-gauge** (a canister with a funnel opening, standing on open ground). The accumulated precipitation (if any) is measured daily in a measuring flask. Daily totals are added to give monthly and annual totals. **Isohyets** are lines on maps joining places receiving equal amounts of rainfall.

Pressure This is measured by means of either a **mercury** or an **aneroid barometer**. Pressure is expressed in millibars. Average pressure at sea-level is 1,012 millibars. Pressure is shown on maps by means of **isobars** (lines joining places of equal pressure, the readings having been corrected to sea-level equivalents).

Relative humidity This is measured by **wet and dry bulb thermometers** kept in a Stevenson's screen. When air contains the maximum possible moisture at a given temperature it is said to be saturated. This is the **dew-point** temperature of the air.

Wind speed and direction Wind speed is measured by means of a **cup-anemometer**. The result may be given in km.p.h. or knots or expressed as a Beaufort number. Wind direction is shown by a **wind-vane**. Note that winds are always named according to the direction from which they blow.

Sunshine This is recorded by traces burnt on to a sensitised card by the sun through a glass sphere.

Cloud cover Cloud conditions are assessed visually. Cloud cover is expressed in oktas or eighths. This is the fraction of the sky covered by cloud. Clouds may be divided into high, medium and low types.

Factors influencing temperature

The temperature of any place is determined by a number of factors. These include (1) **latitude**; (2) **distance from the sea**; (3) **altitude**; (4) **ocean currents**; (5) **cloudiness**; (6)

aspect; (7) **local winds**. Be able to discuss the operation of these factors by reference to specific regions.

Rainfall
The amount

The amount of water vapour that an air mass can contain depends upon its temperature. Warm air can hold more water vapour than cold air. Therefore, all processes of rainfall formation involve cooling of the air and the reduction of its temperature to below dew-point (the temperature at which the air is fully saturated). Three types of rainfall may be distinguished.

(1) **Relief or orographic rainfall** When winds meet a mountain barrier they are forced to rise. As the air rises it expands (owing to the reduced pressure) and cools and may reach dew point. Cloud and rain may thus be produced. On the leeward side of the mountains the air is descending and is consequently warmed by compression. Thus, the windward side of mountain ranges is usually far wetter than the leeward side or **rain shadow**.

(2) **Convectional rainfall** Intensive heating of the land causes rising air. As the air rises it expands and cools and may reach dew point. Usually cumulus clouds are produced. Hail and thunder are often associated with convectional storms. The low pressure created by the rising air currents causes an inward movement of air to the centre of the storm area. Areas most affected by this type of rain are those subject to intensive heating, e.g. equatorial areas; continental interiors in summer.

(3) **Frontal or depressional rainfall** A **front** is a surface dividing two air masses of different temperature and humidity characteristics. When two air masses of different types are drawn together, as in a depression, the warm air tends to rise over the denser cold air (the warm front of a depression). Alternatively the cold air will undercut the warm air, thus lifting it off the ground (the cold front). Thus the warm air may be reduced to dew point as a result of cooling by ascent and contact with the cold air.

The weather map

The symbols employed on the weather maps prepared by the Meteorological Office should be learnt (see page 104). The characteristic features of the most common types of weather situation must also be known. These include the following.

(1) **Depression** This is a centre of low pressure indicated on the map by a series of roughly concentric closed isobars. Winds blow around a depression in an anticlockwise direction in the northern hemisphere, cutting slightly across the isobars towards the centre of low pressure. Air of different types is drawn to this centre, and fronts develop between different air masses. In a typical depression the air lying between the warm and cold fronts to the south of the centre is

called the **warm sector**. Along the **warm front** warm air rises over the cold air to give a wide belt of cloud and rain. In the rear of the depression, along the **cold front**, cold air undercuts the warm sector to give a second belt of rain. Since the cold front advances more rapidly than the warm front it is usual for the warm sector to become completely undercut by the cold air and lifted off the ground. The depression is then said to be **occluded**.

(2) **Anticyclone** This is a centre of high pressure. Isobars form a widely spaced, roughly concentric pattern. Winds blow in a clockwise direction in the northern hemisphere, cutting slightly across the isobars as they blow outwards from the centre. Winds are usually light and variable. Anticyclones give calm stable weather with hot sunny days in summer, and cold clear days in winter. In autumn anticyclones are often associated with fogs (**smog** over urban areas).

(3) **Ridge of high pressure** Lying between two depressions and often an extension of an anticyclone, the high-pressure ridge is usually associated with light winds and fine weather.

(4) **Trough of low pressure** Lying between areas of relatively high pressure, the low-pressure trough may be an extension of a depression. The passage of a trough is usually marked by unsettled, rainy weather, often with squally winds.

World climate regions

Pressure and winds

High-pressure belts are found at latitudes 30°N and 30°S; high-pressure zones are also located at the Poles. Low pressure is experienced at the equator and at 60°N and 60°S. The simplified pattern of winds in response to these pressure belts produces the north-east and south-east trade winds in the tropics, the north-west and south-west variable winds in middle latitudes, and the north-east and south polar winds in high latitudes. This simple pattern is modified by two factors: (1) the uneven distribution of continents and oceans, and (2) the seasonal movements of the overhead sun and the resultant migration of the pressure belts.

Figure 42 on page 111 shows the distribution of the world's major climatic regions. The main characteristics of each climatic type should be learnt, together with the associated vegetation.

(1) **Equatorial climate** This is experienced in areas close to the equator, between latitudes 5°N and 5°S. The sun is high throughout the year (actually overhead at the equator on the equinoxes) and temperatures are constantly high. The annual range of temperature is small. Rainfall is heavy and is of the convectional type. Vegetation is of the **rain-forest** type.

(2) **Tropical continental (Sudan) climate** Areas of tropical

continental climate are found on either side of the equatorial climate belt. The sun is overhead in summer and gives high temperatures and convectional rain. Winters are cooler (although still hot) and there is far less rainfall. Typical vegetation consists of **savanna grass** with scattered trees.

(3) **Hot desert climate** This is experienced along the western sides of continents in or near the tropics. Summers are very hot (the sun is overhead at the tropics) and winters are warm or hot. Owing to the lack of cloud the diurnal range of temperature is great. Onshore winds usually cross cold ocean currents and therefore yield little rain. Owing to the lack of rain **drought-resistant plants** with thorny or waxy leaves, deep roots or some means of water storage are typical.

(4) **Tropical maritime climate** Areas of tropical maritime climate are found on the eastern side of continents between latitudes 5° and 23½° (apart from South East Asia). These areas are dominated by onshore trade winds and therefore have all-seasonal rain. The range of temperature is small owing to the high angle of the sun. Natural vegetation consists of **rain forests**.

(5) **Tropical monsoon climate** This is experienced in South East Asia and northern Australia. The climate is one of marked seasonal contrasts. In winter, high pressure develops over Asia and causes dry offshore winds. In summer, heating of the Asian landmass creates low pressure and causes winds to blow onshore. Therefore, there is a seasonal reversal of winds and a marked division of the year into the hot, wet summer season and the hot, dry winter season. The wettest areas have **evergreen forest**; the drier areas **deciduous woodland or grassland**.

(6) **Mediterranean climate** Areas of Mediterranean climate are found along the western sides of continents between latitudes 30° and 40°. The climate is determined by the seasonal movement of the pressure and wind belts. In winter these areas have onshore westerlies which produce warm, wet winters. In summer offshore trade winds or high-pressure calms dominate the climate to give hot, dry summers. Low-rainfall areas have a poor scrub vegetaion variously referred to as **maquis, garrigue**, or **chaparral**. Wetter areas have **mixed or deciduous woodland**.

(7) **Warm temperate east margin climate** This type of climate is experienced in the same latitudes as Mediterranean climate but on the eastern sides of continents. Offshore variables give dry winters, but trade winds are drawn strongly onshore in summer to give heavy rain. A **woodland or forest vegetation** is typical.

(8) **Cool temperate west margin climate** This is experienced along the western sides of continents between

latitudes 40° and 60°. Onshore westerlies and the passage of depressions throughout the year produce all-seasonal rain. Owing to maritime influences summers are cool and winters mild. Natural vegetation consists of **mixed or deciduous woodland**.

(9) **Cool temperate east margin climate** This is experienced along the eastern sides of continents between latitudes 40° and 50°. The westward movement of air masses from the continental interiors produces very cold winters and hot summers. Depressional rainfall occurs at all seasons. **Woodland or forest** vegetation is typical.

(10) **Temperate interior climate** This is experienced in continental interiors in middle latitudes. Owing to the distance from the sea, the climate is one of extremes: summers are hot and winters very cold. Rainfall is light or moderate with a summer maximum. Grassland vegetation is typical. This is known as **prairie, steppe, pampas** or **veld** in various parts of the world.

(11) **Cold temperate climate** This type of climate is experienced in a broad belt stretching across northern Eurasia and North America. Climate is dominated by the high-latitude position. Summers are short and cool, winters cold or very cold. Precipitation varies according to position, being heaviest along the west coasts. Natural vegetation consists of coniferous forest known as **taiga**.

(12) **Arctic or polar climate** This type of climate is confined to the northernmost parts of Eurasia and North America. Summers are short and cool, winters long and extremely cold. Precipitation is usually light. Typical vegetation consists of mosses, lichens and dwarf species of trees, and is referred to as **tundra**.

World population and food supply

World distribution of population

The total world population of $c.$ 3,865 millions (1973 figure) is very unevenly distributed. About 90 per cent of the world total inhabits the northern hemisphere, and about 85 per cent lives in the Old World (Eurasia). Three main concentrations of population may be noted: namely, **South East Asia** (including the Indian sub-continent), **Europe**, and **north-east North America**. The inhabited parts of the earth's surface are referred to by the term **ecumene**, while the uninhabited or sparsely inhabited regions are known as the **non-ecumene**. The distribution of population is influenced by both physical and social and economic factors. These include altitude, relief, climate, soils, natural vegetation, mineral resources, age of settlement, type of economy and political influences.

World population growth

In recent years world population has increased by approximately 70 millions each year. This figure is equivalent to a population larger than that of the UK or about one-third that of the USA. It represents a monthly addition to the world's population equivalent to a city ten times the size of Leeds.

As early as 1798 Thomas **Malthus** drew attention to the fact that population tends to increase at a faster rate than the increase in food supply and the means of subsistence. He suggested that population has a tendency to increase in a geometric ratio, while food production tends to increase in an arithmetical ratio. Although the ideas of Malthus require modification in view of the technological and medical developments since his time, the basic problem to which he drew attention has never been more urgent than at the present time. Population has already doubled during the present century, and if present rates are maintained will be multiplied sixfold within 100 years.

Yet even now about half the world's population suffers from **malnutrition**, and regional **famine** disasters are commonplace. Malnutrition impairs health and vitality and in turn produces **deficiency diseases**, as well as lowering resistance to other diseases.

Possibilities of increasing world food production

The means of increasing world food supply include the following:

(1) **expansion of agriculture into new lands**, which might involve the introduction of commercial ranching into tropical grasslands, large-scale irrigation schemes, land reclamation, etc.;

(2) **intensification of agriculture in existing farmlands**, which can be achieved by the greater use of fertilisers, the development of higher-yielding strains of seeds, selective stock-breeding, greater use of pesticides, increased mechanisation, programmes of education, improvements in food storage and distribution, etc.; greater use could be made of fish-farming and marine resources, and new sources of protein could be developed.

Agriculture

The type of agriculture practised in any area is the result of many influences. **Physical factors** (climate, soil and relief) determine the broad characteristics of the agricultural pattern, but in most environments there is a wide range of crops and livestock able to thrive under the conditions present. Thus, the actual selection present in the agricultural system is the result of **economic and social influences**, either past or present, such as farm size, system of land tenure, market demand, availability of labour, transport facilities, tariffs and other government influences, etc.

Farming systems
A basic distinction may be drawn between **subsistence** and **commercial** agriculture. Various systems of farming organisation may be noted. These include: (1) **shifting agriculture**; (2) **nomadic herding**; (3) **tropical peasant agriculture**; (4) **plantation agriculture**; (5) **intensive mixed farming**; (6) **market gardening**; (7) **extensive arable farming**; (8) **extensive ranching**. The main characteristics of each type should be learnt together with regional examples.

Crops and livestock
Another approach to the study of agriculture is to examine the distribution of specific types of crops and livestock. In this approach reference is normally made to the physical requirements and tolerances of each crop, its main production areas, volume of production and importance in world trade. This type of information should be learnt for the following crops and livestock: **tropical crops**, rubber, coffee, cocoa, tea, sugar-cane, cotton and rice; **middle-latitude (temperate) crops and livestock**, wheat, barley, oats, rye, maize, potatoes, sugar-beet, dairy and beef cattle and sheep.

Farming problems and techniques
Soil erosion This is the destruction and removal of soil by either wind or water action. Causes include over-cropping and over-grazing, ploughing up slopes, deforestation, etc. Soil erosion is a serious world problem and all continents are affected, e.g. over half the land in the USA is affected in varying degrees. Techniques of prevention and cure include strip-cropping, terracing of hillsides, greater use of fertilisers, contour ploughing, afforestation, diversification of farming, planting of wind-breaks, and education by soil scientists.

Land drainage Techniques of land drainage may be used to increase the agricultural potential of low-lying, water-logged land. These include the digging of ditches, installation of tile- and pipe-drains in fields, straightening of river courses, construction of levées and the building of sluices and pumping stations, etc.

Soil improvements Techniques of soil improvement include the application of fertilisers, and the use of **crop rotations**. In low-rainfall areas various techniques of **dry farming** may be used to conserve soil moisture, e.g. cropping in alternate years.

Irrigation Irrigation is generally practised in areas with less than 500 mm rainfall per year. Methods include rudimentary water wheels, tank storage, artesian wells, basin irrigation and large, modern, multi-purpose schemes with huge river-dams and reservoirs.

Fishing and forestry

Fishing

The chief fishing grounds of the world are found in the shallow waters of continental-shelf areas, especially in high and middle latitudes. Here **plankton** is abundant, e.g. North Sea, White Sea, Grand Banks, etc. Methods of fishing depend to a large extent on whether **pelagic** or **demersal** fish are being sought. Methods include **drifting**, **trawling**, **long-line fishing** and **seine netting**.

Forestry

In many parts of the world the natural forest cover has been seriously depleted by over-cutting of timber. In many countries large-scale **afforestation** schemes have been embarked upon in order to conserve and replenish resources. On a world scale the working of three main types of timber may be noted: namely, **tropical hardwoods** (e.g. mahogany, ebony, teak, etc.), **temperate hardwoods** (e.g. oak, ash, kari, redwood, etc.) and **temperate softwoods** (e.g. pine, spruce, fir, etc.).

Mining

The occurrence of minerals (excluding fuel minerals)

(1) **Bedded ores** These were formed as a result of deposition in former lakes or seas, evaporation from former shallow seas, or the decomposition and partial solution of surface deposits leaving a residual mass. (2) **Veins and lodes** These are related to igneous intrusions. They follow irregular courses through the country rock and are therefore difficult to work. The mineral content often changes with depth. (3) **Alluvial or placer deposits** These are the result of the erosion of either bedded or vein deposits by rivers and the redeposition of the eroded material as sediment.

Exploitation of mineral deposits

The exploitation of mineral deposits depends on a variety of factors other than the mere presence of minerals. The richest and most accessible ores tend to be worked first, and production costs tend to rise with the age of a particular mine. Deposits are rarely worked to complete exhaustion. Many factors determine (1) whether a deposit is initially worked, and (2) when working ceases. These factors include the

quality of the ore, its **accessibility, ease of mining**, the **size** of the deposit, **market demand, technological considerations, labour-supply, water-supply, capital** reserves, **level of economic development**, etc. Methods of mining vary according to the type of mineral and the form in which it occurs. Methods include **shaft mining, adit mining**, and **opencast mining** or quarrying.

Many minerals require processing or concentration at or near the mine prior to transportation to market areas. This generally involves the separation of valuable minerals from the **gangue** or waste material. Of particular importance in this connection is the **flotation process**.

Candidates should be able to provide information about the form of occurrence, methods of mining, chief producers and world trade for the chief mineral products, notably iron, copper, bauxite, tin, lead, zinc, nickel, gold, silver, and various chemical raw materials (salt, sulphur, potash, nitrate and phosphate).

Fuel and power

A close correlation exists between levels of energy consumption and levels of economic development and standards of living. A shortage of fuel and power greatly hinders economic development. The term **energy mix** is used to denote the combination of different types of fuel and power which an economy uses. The exploitation of fuel and power resources in any area depends upon the size of reserves, ease of working, access to markets, labour and capital availability, etc. The earliest forms of power included water and wind. The invention of the steam engine marked the beginning of the modern industrial era. Present power sources include coal, oil (petroleum), natural gas, hydro-electricity (HEP) and nuclear power. Power is important for industry, transport and domestic use.

Coal Coal is a sedimentary rock formed by the partial decomposition of vegetation laid down in swamps and deltas, especially during the Carboniferous period. Coal consists of carbon, hydro-carbons, moisture and non-combustible material. These occur in different proportions in different types of coal. Note the various types of coal, ranging from lignite to anthracite. Different methods of mining are used according to the nature and location of the seams. These include shaft mining, adit mining and opencast mining. Coal is important as a fuel (heating), as a source of power, as a raw material, and for the production of industrial coke. Chief producers include the USA, USSR and China.

Oil Oil is composed of hydro-carbons formed from decaying organic matter sealed in marine sediments during their deposition. Oil collects in porous rocks where conditions of structure have provided the necessary traps. A capping of impervious rock is needed. Be able to draw diagrams showing the various types of oil traps. Exploitation requires vast capital resources. The movement of oil to areas of demand may be by pipelines or oil-tankers. Refining now tends to be carried out in market areas rather than supply areas. Note the wide range of products from modern refineries (petrol, paraffin, diesel oil, bitumen, lubricants, plastics, synthetic rubber, etc.). Chief producers include the USA, USSR, Saudi Arabia, Iran, Venezuela and Kuwait.

Natural gas This is often found in association with oil. Major producers include the USA, USSR, Canada, Venezuela, Netherlands and the UK. At present natural gas supplies 15 per cent of world energy.

Electricity Electricity can be generated from coal, natural gas, oil, nuclear fuel and water. The choice of fuel depends on the availability and relative costs of different fuels, security of supply, etc. Large-scale **HEP developments** ideally require rivers with a large discharge, rapid flow, regular régime, and lakes along their courses to regulate flow and filter sediment. Installation costs are great, but once completed, the cost of HEP is relatively low.

Transport facilities relating to fuel and power can be divided into two groups; **discontinuous** forms such as road, rail, canal and sea, and **continuous** forms such as pipelines and transmission lines. In general discontinuous forms have low capital costs and high running costs, while continuous forms have high capital costs and low running costs.

Industry

Industrial activities may be divided into **primary** industry (mining, fishing, forestry and agriculture), **secondary** industry, (processing and manufacturing) and **tertiary** industry (service industries such as transport, marketing, finance, etc.). We are concerned here with secondary industry. Within the manufacturing sector it is possible to distinguish between the production of **capital** goods and the production of **consumer** goods.

The location of industry

A variety of factors influences the location of any industrial site. Precise assessment of the factors involved is difficult. Among the most important factors involved are the following: (1) **fuel and raw materials**; (2) **transport facilities**; (3) **labour-supply** (the size and skills of the labour force and the cost of labour); (4) **capital**; (5) **market demand** (this involves consideration of the size and affluence of the market and its accessibility); (6) **government influences** (the influence of subsidies, tax concessions and industrial relocation policies, etc.); (7) **industrial concentration** (as an area develops industrially there is a tendency for other industries to be attracted to it).

There is a tendency for an industry, once established in an area, to remain there, even if the original locational factors no longer apply (e.g. the supply of raw materials). This is termed **geographical or industrial inertia**. Therefore, it is often difficult to explain the location of industry simply in terms of present-day factors.

Major manufacturing industries
Iron and steel The requirements are iron ore, coke, limestone, water, cheap transport and a large labour force (in part highly skilled). Plants may be sited near the power source or near the iron (or the entry port if imported) or between the two. Initial processing is the conversion of the iron ore to pig iron in a blast furnace. Slag (a by-product of this process) may be used for fertiliser. The pig iron is then further processed to produce steel. This involves reheating and the addition of carbon, scrap iron and small amounts of other metals (nickel, chromium, tungsten, etc.) according to the type of steel being produced. Major producers of steel include the USSR, USA, Japan, West Germany and the UK. A large part of the production is used by **engineering industries** for a variety of products, e.g. machine tools, agricultural implements, cars, ships, etc.

Shipbuilding Requirements include sheltered sea-water estuaries, local or nearby iron and steel, power supply, products of marine engineering industries (e.g. turbines), timber for fittings, and a large labour-supply (much of it skilled). Chief shipbuilding nations are Japan, Sweden, West Germany, UK, Spain, France and USA.

Motor industry This is essentially an assembly industry which uses components from a variety of suppliers. Considerable flexibility of location is therefore possible. Availability of capital, large labour-supply and access to markets or ports for exports are all important factors. The rapid growth of the industry and its concentration in the hands of a few giant companies are notable characteristics.

Radio and electronics Like the motor industry this is often an assembly industry. The industry tends to be attracted to market areas as well as research establishments where new ideas and research personnel are available.

Chemicals The products of the chemical industry are

221

extremely diverse, e.g. **industrial chemicals** (acids, alkalis, dyes, bleaches, etc.), **commercial fertilisers** (nitrates, phosphates, etc.), **plastics, synthetic textile fibres**, etc. The industry may be sited on coalfields, near oil-refineries, on salt deposits, near ports if imported raw materials are used, or near centres of demand.

Textiles The production of textiles is one of the world's leading industrial activities. Production includes cotton, wool, linen and synthetic textile fibres. Factors of location are varied and complex, and geographical inertia is frequently evident. Most of the major producers of textiles lie in the northern hemisphere, e.g. USA, UK, Belgium, France, Italy, etc.

Industry and underdeveloped countries
Most underdeveloped countries regard industrialisation as the means of stimulating economic development. However, large-scale industrial development is difficult owing to **problems of labour-supply, lack of capital** and **marketing difficulties**.

Settlement

Rural settlement
Rural settlements vary in size from single homesteads and farms, through hamlets and villages (houses, shops and services) to small market towns (limited in their range of functions).

A basic distinction can be made between **dispersed** and **nucleated** settlement. In the former type houses may be scattered throughout an area (e.g. in parts of the Fens), while in the latter type the buildings are closely clustered around a church or village green or some other central point.

Various types of village may be noted: (1) **spring-line villages**, which grew up where springs provided a source of water-supply, especially at the foot of scarp slopes; (2) **flood-plain villages**, which are usually found along valley sides away from areas of flood risk, and have the advantages of fertile soils, shelter, and local water-supply; (3) **hill-top villages**, the original location of which was often chosen for defensive purposes or in some cases because the hill-top was a dry site in a badly drained area; (4) **bridge-point villages**, which controlled river-crossing places in early times; (5) **street villages**, linear settlements which grew up along lines of communication; (6) **cruciform villages**, which developed at an intersection of routes; (7) **green villages**, which grew up around a central green or open space.
Learn specific examples of each type from OS maps.

Urban settlement

Towns may be classified either according to their **function** or the nature of their **site**. On the basis of function the following types of town may be identified: (1) commercial centres; (2) mining towns; (3) industrial towns; (4) administrative centres; (5) cultural and educational centres; (6) religious centres; (7) tourist centres and holiday resorts; (8) residential towns (dormitory towns); (9) new towns; (10) transport and route centres; (11) ports (commercial ports, entrepôts, packet-stations, fishing ports, tanker ports, outports, ports of call and naval ports). On the basis of site the following types may be identified: (1) river sites (gap towns, meander-core towns, bridge-point towns, confluence towns, delta towns, head of navigation towns); (2) coastal sites; (3) lakeside sites; (4) sites on alluvial plains; (5) defensive sites; (6) frontier sites, etc.

Urban structure

The internal structure of towns forms an interesting area of study. Field- or map-study of urban plans will often reveal distinctive areas of growth or land use within the city. The following urban zones can usually be distinguished: (1) **central business districts** (CBDs); (2) **industrial areas**; (3) **residential districts**; (4) **open spaces**.

Transport

One of the bases of economic activity is that of trade between different areas producing different goods, and transport is of vital importance as the means by which this trade takes place. The type and quality of the transport system in any area varies according to physical, economic and historical factors.

Roads These vary in quality from dirt trackways to motorways costing £1 million per kilometre to construct. Roads provide a rapid, door-to-door service for goods, i.e. a single operation in the movement of goods between producer and consumer. Roads are less restricted by relief than railways, but are more affected by adverse weather conditions.

Railways These cannot run on steep gradients, so construction may involve tunnelling or the building of embankments and cuttings. Therefore initial construction is costly. Motive power may come from steam, diesel oil or electricity. Railways enable the fast movement of bulk commodities as well as passengers and mail. Railways are important in densely-populated and industrialised countries as well as in the opening up of new, developing areas.

Inland waterways These provide a cheap but slow method of transport for bulky, low-value commodities. The use of waterways has declined in many countries in favour of quicker, more efficient forms of transportation. Problems of water transport include fluctuations of water-level, shallows, falls, winter freezing, and the restricted coverage of the waterways network.

Ocean shipping This is a slow but relatively cheap method of moving bulky goods. The types of ship involved vary from small coasters to the latest 300,000-tonne bulk-carriers and tankers. Leading shipping nations include Liberia, Japan, UK, USA, Greece and Norway.

Air transport Air transport has the great advantage of high speed (passengers, mail, etc.), but cargoes are limited to low-bulk, expensive items which can withstand the high charges involved. In recent years commercial air services have captured a great deal of traffic from shipping lines. This is especially true of the transatlantic route.